Springer Optimization and Its Applications

Volume 67

Aims and Scope

Optimization has continued to expand in all directions at an astonishing rate. New algorithmic and theoretical techniques are continually developing and the diffusion into other disciplines is proceeding at a rapid pace, with a spot light on machine learning, artificial intelligence, and quantum computing. Our knowledge of all aspects of the field has grown even more profound. At the same time, one of the most striking trends in optimization is the constantly increasing emphasis on the interdisciplinary nature of the field. Optimization has been a basic tool in areas not limited to applied mathematics, engineering, medicine, economics, computer science, operations research, and other sciences.

The series **Springer Optimization and Its Applications (SOIA)** aims to publish state-of-the-art expository works (monographs, contributed volumes, textbooks, handbooks) that focus on theory, methods, and applications of optimization. Topics covered include, but are not limited to, nonlinear optimization, combinatorial optimization, continuous optimization, stochastic optimization, Bayesian optimization, optimal control, discrete optimization, multi-objective optimization, and more. New to the series portfolio include Works at the intersection of optimization and machine learning, artificial intelligence, and quantum computing.

Volumes from this series are indexed by Web of Science, zbMATH, Mathematical Reviews, and SCOPUS.

More information about this series at http://www.springer.com/series/7393

Michael L. Bynum • Gabriel A. Hackebeil
William E. Hart • Carl D. Laird
Bethany L. Nicholson • John D. Siirola
Jean-Paul Watson • David L. Woodruff

Pyomo – Optimization Modeling in Python

Third Edition

 Springer

Michael L. Bynum
Sandia National Laboratories
Albuquerque, NM, USA

Gabriel A. Hackebeil
Deepfield Nokia
Ann Arbor, MI, USA

William E. Hart
Sandia National Laboratories
Albuquerque, NM, USA

Carl D. Laird
Sandia National Laboratories
Albuquerque, NM, USA

Bethany L. Nicholson
Sandia National Laboratories
Albuquerque, NM, USA

John D. Siirola
Sandia National Laboratories
Albuquerque, NM, USA

Jean-Paul Watson
Lawrence Livermore National Laboratory
Livermore, CA, USA

David L. Woodruff
Graduate School of Management
University of California
Davis, CA, USA

ISSN 1931-6828   ISSN 1931-6836 (electronic)
Springer Optimization and Its Applications
ISBN 978-3-030-68930-8   ISBN 978-3-030-68928-5 (eBook)
https://doi.org/10.1007/978-3-030-68928-5

This Springer imprint is published by the registered company Springer Nature Switzerland AG
The registered company address is: Gewerbestrasse 11, 6330 Cham, Switzerland

To Pyomo users, past, present, and future.

Preface

This book describes a tool for mathematical modeling: the Python Optimization Modeling Objects (Pyomo) software. Pyomo supports the formulation and analysis of mathematical models for complex optimization applications. This capability is commonly associated with algebraic modeling languages (AMLs), which support the description and analysis of mathematical models with a high-level language. Although most AMLs are implemented in custom modeling languages, Pyomo's modeling objects are embedded within Python, a full-featured high-level programming language that contains a rich set of supporting libraries. Pyomo has won awards from the R&D100 organization and from the INFORMS Computing Society.

Modeling is a fundamental process in many aspects of scientific research, engineering and business, and the widespread availability of computing has made the numerical analysis of mathematical models a commonplace activity. Furthermore, AMLs have emerged as a key capability for robustly formulating large models for complex, real-world applications [37]. AMLs streamline the process of formulating models by simplifying the management of sparse data and supporting the natural expression of model components. Additionally, AMLs like Pyomo support scripting with model objects, which facilitates the custom analysis of complex problems.

The core of Pyomo is an object-oriented capability for representing optimization models. Pyomo also contains packages that define modeling extensions and model reformulations. Pyomo also includes packages that define interfaces to solvers like CPLEX and Gurobi, as well as solver services like NEOS.

Goals of the Book

This third edition provides an updated description of Pyomo's modeling capabilities. A key goal of this book is to provide a broad description of Pyomo that will enable the user to develop and optimize models with Pyomo. The book uses many examples to illustrate different techniques that can be used to formulate models.

Another goal of this book is to illustrate the breadth of Pyomo's capabilities. Pyomo supports the formulation and analysis of common optimization models, including linear programs, mixed-integer linear programs, nonlinear programs, mixed-integer nonlinear programs, mathematical programs with equilibrium constraints, constraints and objectives based on differential equations, generalized disjunctive programs, among others. Additionally, Pyomo includes solver interfaces for a variety of widely used optimization software packages, including CBC, CPLEX, GLPK, and Gurobi. Additionally, Pyomo models can be optimized with optimizers like IPOPT that employ the AMPL Solver Library interface.

Finally, a goal of this book is to help users get started with Pyomo even if they have little knowledge of Python. Appendix A provides a quick introduction to Python, but we have been impressed with how well Python reference texts support new Pyomo users. Although Pyomo introduces Python objects and a process for applying them, the expression of models with Pyomo strongly reflects Python's clean, concise syntax.

However, our discussion of Pyomo's advanced modeling capabilities assumes some background in object-oriented design and features of the Python programming language. For example, our discussion of modeling components distinguishes between class definitions and class instances. We have not attempted to describe these advanced features of Python in the book. Thus, a user should expect to develop some familiarity with Python in order to effectively understand and use advanced modeling features.

Who Should Read This Book

This book provides a reference for students, academic researchers and practitioners. The design of Pyomo is simple enough that it has been effectively used in the classroom with undergraduate and graduate students. However, we assume that the reader is generally familiar with optimization and mathematical modeling. Although this book does not contain a glossary, we recommend the Mathematical Programming Glossary [32] as a reference for the reader.

Pyomo is also a valuable tool for academic researchers and practitioners. A key focus of Pyomo development has been on the ability to support the formulation and analysis of real-world applications. Consequently, issues like run-time performance and robust solver interfaces are important.

Additionally, we believe that researchers will find that Pyomo provides an effective framework for developing high-level optimization and analysis tools. For example, Pyomo is the basis of a package for optimization under uncertainty called mpi-sppy, and it leverages the fact that Pyomo's modeling objects are embedded within a full-featured high-level programming language. This allows for transparent parallelization of sub-problems using Python parallel communication libraries. This ability to support generic solvers for complex models is very powerful, and we believe that it can be used with many other optimization analysis techniques.

Revisions for the Third Edition

We have made several major changes while preparing the third edition of this book. A subtle change that permeates the book is in how we recommend that Pyomo be imported. A bigger change that permeates the book is an emphasis on concrete models. The introductory chapter starts with a concrete model, and we emphasize concrete models in most chapters other than the chapter devoted entirely to abstract models. This does not reflect a change in Pyomo's capabilities, but rather a recognition that concrete models provide fewer restrictions on the specification and use of Pyomo models. For example, data can be loaded by the user using general Python utilities rather than the mechanisms supported specifically for abstract models. Thus, concrete models enable a more general discussion of Pyomo's potential. Finally, we have reorganized much of the material, added new examples, and added a chapter on how modelers can improve the performance of their models. Two chapters have been dropped from the book: the chapter on bilevel programming and the chapter on PySP, which offered support for optimization under uncertainty. These capabilities are available in other ways and are no longer included in Pyomo software releases.

Acknowledgments

We are grateful for the efforts of many people who have supported current and previous editions of this book. We thank Elizabeth Loew at Springer for helping shepherd this book from an initial concept to final production; her enthusiasm for publishing is contagious. Also, we thank Madelynne Farber at Sandia National Laboratories for her guidance with the legal process for releasing open source software and book publishing. Finally, we thank Doug Prout for developing the Pyomo, PySP, Pyomo.DAE, and Coopr logos.

We are indebted to those who spent time and effort reviewing this book. Without them, it would contain many typos and software bugs. So, thanks to Jack Ingalls, Zev Friedman, Harvey Greenberg, Sean Legg, Angelica Wong, Daniel Word, Deanna Garcia, Ellis Ozakyol, and Florian Mader. Special thanks to Amber Gray-Fenner and Randy Brost.

We are particularly grateful to the growing community of Pyomo users. Your interest and enthusiasm for Pyomo was the most important factor in our decision to write this book. We thank the early adopters of Pyomo who have provided detailed feedback on the design and utility of the software, including Fernando Badilla, Steven Chen, Ned Dmitrov, YueYue Fan, Eric Haung, Allen Holder, Andres Iroume, Darryl Melander, Carol Meyers, Pierre Nancel-Penard, Mehul Rangwala, Eva Worminghaus and David Alderson. Your feedback continues to have a major impact on the design and capabilities of Pyomo.

We also thank our friends in the COIN-OR project for supporting the Pyomo software. Although the main development site for Pyomo is hosted at GitHub, our partnership with COIN-OR is a key part of our strategy to ensure that Pyomo remains a viable open source software project.

A special thanks goes to our collaborators who have contributed to packages in Pyomo: Francisco Muñoz, Timothy Ekl, Kevin Hunter, Patrick Steele, and Daniel Word. We also thank Tom Brounstein, Dave Gay, and Nick Benevidas for helping develop Python modules and documentation for Pyomo.

The authors gratefully acknowledge this support that contributed to the development of this book: National Science Foundation under Grant CBET#0941313 and CBET#0955205, the Office of Advanced Scientific Computing Research within the U.S. Department of Energy Office of Science, the U.S. Department of Energy ARPA-E under the Green Electricity Network Integration program, the Institute for the Design of Advanced Energy Systems (IDAES) with funding from the Simulation-Based Engineering, Crosscutting Research Program within the U.S. Department of Energy's Office of Fossil Energy, the U.S. Department of Energy's Office of Electricity Advanced Grid Modeling (AGM) program, and the Laboratory Directed Research and Development program at Sandia National Laboratories

And finally, we would like to thank our families and friends for putting up with our passion for optimization software.

Disclaimers

Livermore National Security, LLC, nor any of their employees makes any warranty, expressed or implied, or assumes any legal liability or responsibility for the accuracy, completeness, or usefulness of any information, apparatus, product, or process disclosed, or represents that its use would not infringe privately owned rights. Reference herein to any specific commercial product, process, or service by trade name, trademark, manufacturer, or otherwise does not necessarily constitute or imply its endorsement, recommendation, or favoring by the United States government or Lawrence Livermore National Security, LLC. The views and opinions of authors expressed herein do not necessarily state or reflect those of the United States government or Lawrence Livermore National Security, LLC, and shall not be used for advertising or product endorsement purposes. This work was performed in part under the auspices of the U.S. Department of Energy by Lawrence Livermore National Laboratory under Contract DE-AC52-07NA27344.

Comments and Questions

This book documents the capabilities of the Pyomo 6.0 release. Further information is available on the Pyomo website, including errata:

```
http://www.pyomo.org
```

Pyomo's open source software is hosted at GitHub, and the examples used in this book are included in the `pyomo/examples/doc/pyomobook` directory:

```
https://github.com/Pyomo/pyomo
```

Many Pyomo questions are posed and answered on Stack Overflow:

```
https://stackoverflow.com/
```

We encourage feedback from readers, either through direct communication with the authors or with the Pyomo Forum:

```
pyomo-forum@googlegroups.com
```

Good luck!

Albuquerque, New Mexico, USA   *Michael Bynum*
Ann Arbor, Michigan, USA   *Gabe Hackebeil*
Albuquerque, New Mexico, USA   *William Hart*
Albuquerque, New Mexico, USA   *Carl Laird*
Albuquerque, New Mexico, USA   *Bethany Nicholson*
Albuquerque, New Mexico, USA   *John Siirola*
Livermore, California, USA   *Jean-Paul Watson*
Davis, California, USA   *David Woodruff*
                                                                 5 January, 2021

Contents

Part II Advanced Topics   89

Chapter 1
Introduction

Abstract This chapter introduces and motivates Pyomo, a Python-based tool for
modeling and solving optimization problems. Modeling is a fundamental process in
many aspects of scientific research, engineering, and business. Algebraic modeling
languages like Pyomo are high-level languages for specifying and solving mathematical optimization problems. Pyomo is a flexible, extensible modeling framework that captures and extends central ideas found in modern algebraic modeling
languages, all within the context of a widely used programming language.

1.1 Modeling Languages for Optimization

This book describes a tool for mathematical modeling: the Python Optimization
Modeling Objects (Pyomo) software package. Pyomo supports the formulation and
analysis of mathematical models for complex optimization applications. This capability is commonly associated with commercial algebraic modeling languages
(AMLs) such as AIMMS [1], AMPL [2], and GAMS [22]. Pyomo implements a
rich set of modeling and analysis capabilities, and it provides access to these capabilities within Python, a full-featured, high-level programming language with a
large set of supporting libraries.

Optimization models define the goals or objectives for a system under consideration. Optimization models can be used to explore trade-offs between goals and
objectives, identify extreme states and worst-case scenarios, and identify key factors
that influence phenomena in a system. Consequently, optimization models are used
to analyze a wide range of scientific, business, and engineering applications.

The widespread availability of computing resources has made the numerical analysis of optimization models commonplace. The computational analysis of an optimization model requires the specification of a model that is communicated to a
solver software package. Without a language to specify optimization models, the
process of writing input files, executing a solver, and extracting results from a solver
is tedious and error-prone. This difficulty is compounded in complex, large-scale,

real-world applications that are difficult to debug when errors occur. Additionally, solvers use many different input formats, but few of them are considered to be standards. Thus, the application of multiple solvers to analyze a single optimization model introduces additional complexities. Furthermore, model verification (i.e., ensuring that the model communicated to the solver accurately reflects the model the developer intended to express) is extremely difficult without high-level languages for expressing models.

AMLs are high-level languages for describing and solving optimization problems [26, 37]. AMLs minimize the difficulties associated with analyzing optimization models by enabling high-level specification of optimization problems. Furthermore, AML software provides rigorous interfaces to external solver packages that are used to analyze problems, and it allows the user to interact with solver results in the context of their high-level model specification.

Custom AMLs like AIMMS [1], AMPL [2, 21], and GAMS [22] implement optimization model specification languages with an intuitive and concise syntax for defining variables, constraints, and objectives. Further, these AMLs support specification of abstract concepts such as sparse sets, indices, and algebraic expressions, which are essential when specifying large-scale, real-world problems with thousands or millions of constraints and variables. These AMLs can represent a wide variety of optimization models, and they interface with a rich set of solver packages. AMLs are increasingly being extended to include custom scripting capabilities, which enables expression of high-level analysis algorithms concurrently with optimization model specifications.

A complementary strategy is to use an AML that extends a *standard* high-level programming language (as opposed to being based a proprietary language) to formulate optimization models that are analyzed with solvers written in low-level languages. This two-language approach leverages the flexibility of the high-level language for formulating optimization problems and the efficiency of the low-level language for numerical computations. This is an increasingly common approach for scientific computing software. The Matlab TOMLAB Optimization Environment [57] is among the most mature optimization software package using this approach; Pyomo strongly leverages this approach as well. Similarly, standard programming languages like Java and C++ have been extended to include AML constructs. For example, modeling libraries like FlopC++ [19], OptimJ [47], and JuMP [13] support the specification of optimization models using an object-oriented design in C++, Java, and Julia, respectively. Although these modeling libraries sacrifice some of the intuitive mathematical syntax of a custom AML, they allow the user to leverage the flexibility of modern high-level programming languages. A further advantage of these AML libraries is that they can link directly to high-performance optimization libraries and solvers, which can be an important consideration in some applications.

1.2 Modeling with Pyomo

The goal of Pyomo is to provide a platform for specifying optimization models that embodies central ideas found in modern AMLs, within a framework that promotes flexibility, extensibility, portability, openness, and maintainability. Pyomo is an AML that extends Python to include objects for optimization modeling [30]. These objects can be used to specify optimization models and translate them into various formats that can be processed by external solvers.

We now provide some motivating examples to illustrate the use of Pyomo in specifying optimization models.

1.2.1 Simple Examples

Consider the following linear program (LP):

$$\min x_1 + 2x_2$$
$$\text{s.t.} \quad 3x_1 + 4x_2 \geq 1$$
$$2x_1 + 5x_2 \geq 2$$
$$x_1, x_2 \geq 0$$

This LP can be easily expressed in Pyomo as follows:

```
import pyomo.environ as pyo

model = pyo.ConcreteModel()
model.x_1 = pyo.Var(within=pyo.NonNegativeReals)
model.x_2 = pyo.Var(within=pyo.NonNegativeReals)
model.obj = pyo.Objective(expr=model.x_1 + 2*model.x_2)
model.con1 = pyo.Constraint(expr=3*model.x_1 + 4*model.x_2 >= 1)
model.con2 = pyo.Constraint(expr=2*model.x_1 + 5*model.x_2 >= 2)
```

The first line is a standard Python import statement that initializes the Pyomo environment and loads Pyomo's core modeling component library. The next lines construct a model object and define model attributes. This example describes a *concrete* model. Model components are objects that are attributes of a model object, and the `ConcreteModel` object initializes each model component as they are added. The model decision variables, constraints, and objective are defined using Pyomo *model components*.

Users rarely have a single instance of a particular optimization problem to solve. Rather, they commonly have a general optimization model and then create a particular instance of that model using specific data. For example, the following equations represent an LP with scalar parameters n and m, vector parameters b and c, and matrix parameter a:

$$\min \sum_{i=1}^{n} c_i x_i$$
$$\text{s.t.} \quad \sum_{i=1}^{n} a_{ji} x_i \geq b_j \ \forall j = 1 \ldots m$$
$$x_i \geq 0 \qquad \forall i = 1 \ldots n$$

This LP can be expressed with a concrete model in Pyomo as follows:

```python
import pyomo.environ as pyo
import mydata

model = pyo.ConcreteModel()

model.x = pyo.Var(mydata.N, within=pyo.NonNegativeReals)

def obj_rule(model):
    return sum(mydata.c[i]*model.x[i] for i in mydata.N)
model.obj = pyo.Objective(rule=obj_rule)

def con_rule(model, m):
    return sum(mydata.a[m,i]*model.x[i] for i in mydata.N) \
                >= mydata.b[m]
model.con = pyo.Constraint(mydata.M, rule=con_rule)
```

This script requires that the data used to construct the model is available while each modeling component is constructed. In this example, the necessary data exists in `mydata.py`:

```python
N = [1,2]
M = [1,2]
c = {1:1, 2:2}
a = { (1,1):3, (1,2):4, (2,1):2, (2,2):5}
b = {1:1, 2:2}
```

This LP can also be viewed as an *abstract* mathematical model, where unspecified, symbolic parameter values are later defined when the model is initialized. For example, this LP can be expressed as an abstract model in Pyomo as follows:

```python
import pyomo.environ as pyo

model = pyo.AbstractModel()

model.N = pyo.Set()
model.M = pyo.Set()
model.c = pyo.Param(model.N)
model.a = pyo.Param(model.M, model.N)
model.b = pyo.Param(model.M)

model.x = pyo.Var(model.N, within=pyo.NonNegativeReals)

def obj_rule(model):
    return sum(model.c[i]*model.x[i] for i in model.N)
model.obj = pyo.Objective(rule=obj_rule)

def con_rule(model, m):
    return sum(model.a[m,i]*model.x[i] for i in model.N) \
                >= model.b[m]
model.con = pyo.Constraint(model.M, rule=con_rule)
```

This example includes model components that provide abstract or symbolic definitions of set and parameter values. The `AbstractModel` object defers initialization of model components until a *model instance* is created, using user-supplied set and parameter data. Both concrete and abstract models can be initialized with data from a variety of different data sources, including data command files that are adapted from AMPL's data commands. For example:

```
param : N : c :=
1 1
2 2 ;

param : M : b :=
1 1
2 2 ;

param a :=
1 1 3
1 2 4
2 1 2
2 2 5 ;
```

1.2.2 Graph Coloring Example

We further illustrate Pyomo's modeling capabilities with a simple, well-known optimization problem: minimum graph coloring (also known as vertex coloring). The graph coloring problem concerns the assignment of colors to vertices of a graph such that no two adjacent vertices share the same color. Graph coloring has many practical applications, including register allocation in compilers, resource scheduling, and pattern matching, and it appears as a kernel in recreational puzzles like Sudoku.

Let $G = (V, E)$ denote a graph with vertex set V and edge set $E \subseteq V \times V$. Given G, the objective in the minimum graph coloring problem is to find a valid coloring that uses the minimum number of distinct colors. For simplicity, we assume that the edges in E are ordered such that if $(v_1, v_2) \in E$ then $v_1 < v_2$. Let k denote the maximum number of colors, and define the set of possible colors $C = \{1, \ldots, k\}$.

We can represent the minimum graph coloring problem as the following integer program (IP):

$$
\begin{aligned}
\min \quad & y \\
\text{s.t.} \quad & \sum_{c \in C} x_{v,c} = 1 && \forall v \in V \\
& x_{v_1,c} + x_{v_2,c} \le 1 && \forall (v_1, v_2) \in E \\
& y \ge c \cdot x_{v,c} && \forall v \in V, c \in C \\
& x_{v,c} \in \{0, 1\} && \forall v \in V, c \in C
\end{aligned}
\tag{1.1}
$$

In this formulation, the variable $x_{v,c}$ equals one if vertex v is colored with color c and zero otherwise; y denotes the number of colors that are used. The first constraint requires that each vertex is colored with exactly one color. The second constraint requires that vertices that are connected by an edge must have different colors. The third constraint defines a lower bound on y that guarantees that y will be no less than the number of colors used in a solution. The fourth and final constraint enforces the binary constraint on $x_{v,c}$.

Figure 1.1 shows a Pyomo specification of the above graph coloring formulation, using a concrete model; the example is adapted from Gross and Yellen [27]. This specification consists of Python commands that define a `ConcreteModel` object, and then define various attributes of this object, including variables, constraints, and the optimization objective. Lines 10–24 define the model data. Line 28 is a standard Python import statement that adds all of the symbols (e.g., classes and functions) defined in `pyomo.environ` to the current Python namespace. Line 31 specifies creation of the `model` object, which is an instance of the `ConcreteModel` class. Lines 34 and 35 define the model decision variables. Note that y is a scalar variable, while x is a two-dimensional array of variables. The remaining lines in the example define the model constraints and objective. The `Objective` class defines a single optimization objective using the `expr` keyword option. The `ConstraintList` class defines a list of constraints, which are individually added.

When compared to custom AMLs, Pyomo models are clearly more verbose (e.g., see Hart et al. [30]). However, this example illustrates how Python's clean syntax still allows Pyomo to express mathematical concepts intuitively and concisely. Aside from the use of Pyomo classes, this example employs standard Python syntax and methods. For example, line 41 uses Python's generator syntax to iterate over all elements of the `colors` set and apply the Python `sum` function to the result. Although Pyomo does include some utility functions to simplify the construction of expressions, Pyomo does not rely on sophisticated extensions of core Python functionality.

1.2.3 Key Pyomo Features

Python

Python's clean syntax enables Pyomo to express mathematical concepts in an intuitive and concise manner. Furthermore, Python's expressive programming environment can be used to formulate complex models and to define high-level solvers that customize the execution of high-performance optimization libraries. Python provides extensive scripting capabilities, allowing users to analyze Pyomo models and solutions, leveraging Python's rich set of third-party libraries (e.g., numpy, scipy, and matplotlib). Finally, the embedding of Pyomo in Python allows users to learn core syntax through Python's rich documentation.

Customizable Capability

Pyomo is designed to support a "stone soup" development model in which each developer "scratches their own itch." A key element of this design is the plug-in framework that Pyomo uses to integrate model components, model transformations,

```
 1   #
 2   # Graph coloring example adapted from
 3   #
 4   #   Jonathan L. Gross and Jay Yellen,
 5   #   "Graph Theory and Its Applications, 2nd Edition",
 6   #   Chapman & Hall/CRC, Boca Raon, FL, 2006.
 7   #
 8
 9   # Define data for the graph of interest.
10   vertices = set(['Ar', 'Bo', 'Br', 'Ch', 'Co', 'Ec',
11                   'FG', 'Gu', 'Pa', 'Pe', 'Su', 'Ur', 'Ve'])
12
13   edges = set([('FG','Su'), ('FG','Br'), ('Su','Gu'),
14                ('Su','Br'), ('Gu','Ve'), ('Gu','Br'),
15                ('Ve','Co'), ('Ve','Br'), ('Co','Ec'),
16                ('Co','Pe'), ('Co','Br'), ('Ec','Pe'),
17                ('Pe','Ch'), ('Pe','Bo'), ('Pe','Br'),
18                ('Ch','Ar'), ('Ch','Bo'), ('Ar','Ur'),
19                ('Ar','Br'), ('Ar','Pa'), ('Ar','Bo'),
20                ('Ur','Br'), ('Bo','Pa'), ('Bo','Br'),
21                ('Pa','Br')])
22
23   ncolors = 4
24   colors = range(1, ncolors+1)
25
26
27   # Python import statement
28   import pyomo.environ as pyo
29
30   # Create a Pyomo model object
31   model = pyo.ConcreteModel()
32
33   # Define model variables
34   model.x = pyo.Var(vertices, colors, within=pyo.Binary)
35   model.y = pyo.Var()
36
37   # Each node is colored with one color
38   model.node_coloring = pyo.ConstraintList()
39   for v in vertices:
40       model.node_coloring.add(
41               sum(model.x[v,c] for c in colors) == 1)
42
43   # Nodes that share an edge cannot be colored the same
44   model.edge_coloring = pyo.ConstraintList()
45   for v,w in edges:
46       for c in colors:
47           model.edge_coloring.add(
48               model.x[v,c] + model.x[w,c] <= 1)
49
50   # Provide a lower bound on the minimum number of colors
51   # that are needed
52   model.min_coloring = pyo.ConstraintList()
53   for v in vertices:
54       for c in colors:
55           model.min_coloring.add(
56               model.y >= c * model.x[v,c])
57
58   # Minimize the number of colors that are needed
59   model.obj = pyo.Objective(expr=model.y)
```

Fig. 1.1: A concrete Pyomo model for the minimum graph coloring problem.

solvers, and solver managers. A plug-in framework manages the registration of these capabilities. Thus, users can customize Pyomo in a modular manner without the risk of destabilizing core functionality.

Command-Line Tools and Scripting

Pyomo models can be analyzed both using command-line tools and via Python scripts. The `pyomo` command line utility provides a generic interface to most Pyomo modeling capabilities. The `pyomo` command supports a generic optimization process. This process can easily be replicated in a Python script and further customized for a user's specific needs.

Concrete and Abstract Model Definitions

The examples in Section 1.2.1 illustrate Pyomo's support for both concrete and abstract model definitions. The difference between these modeling approaches relates to when modeling components are initialized: concrete models immediately initialize components, and abstract models delay the initialization of components until a later model initialization action. Consequently, these modeling approaches are equivalent, and the choice of approach depends on the context in which Pyomo is used and user preference. Both types of models can be easily initialized with data from a wide range of data sources (e.g., csv, json, yaml, excel, and databases).

Object-Oriented Design

Pyomo employs an object-oriented library design. Models are Python objects, and model components are attributes of these models. This design allows Pyomo to automatically manage the naming of modeling components, and it naturally segregates modeling components within different model objects. Pyomo models can be further structured with blocks, which support a hierarchical nesting of model components. Many of Pyomo's advanced modeling features leverage this structured modeling capability.

Expressive Modeling Capability

Pyomo's modeling components can be used to express a wide range of optimization problems, including but not limited to:

- linear programs,
- quadratic programs,
- nonlinear programs,
- mixed-integer linear programs,

- mixed-integer quadratic programs,
- generalized disjunctive programs,
- mixed-integer stochastic programs,
- dynamic problems with differential algebraic equations, and
- mathematical programs with equilibrium constraints.

Solver Integration

Pyomo supports both tightly and loosely coupled solver interfaces. Tightly coupled modeling tools directly access optimization solver libraries (e.g., via static or dynamic linking), and loosely coupled modeling tools apply external optimization executables (e.g., through the use of system calls). Many optimization solvers read problems from well-known data formats (e.g., the AMPL nl format [24]); these solvers are loosely coupled with Pyomo. Solvers with Python interfaces (e.g., Gurobi and CPLEX) can be tightly coupled, which avoids writing external files.

Open Source

Pyomo is managed as an open source project to facilitate transparency in software design and implementation. Pyomo is licensed under the BSD license [8], which has few restrictions on government or commercial use. Pyomo is managed at GitHub [53], and through the COIN-OR project [9]. Developer and user mailing lists are managed on Google Groups. There is growing evidence that the reliability of open source software is similar to closed source software [3, 59], and Pyomo is carefully managed to ensure the robustness and reliability for users.

1.3 Getting Started

In order to execute all of the examples in this book the following software should be installed:

- Python 3.6 or higher (although almost all examples will work with earlier versions of Python). Pyomo currently relies on CPython; there is only support for Jython and PyPy for a subset of Pyomo's capability.
- Pyomo 6.0, which is used throughout this book.
- The GLPK [25] solver, which is used to generate output for most examples in this book. Other LP and MILP solvers can be used for these examples, but the GLPK software is easily installed and widely available.
- The IPOPT [34] solver, which is used to generate output for nonlinear model examples. Other nonlinear optimizers can be easily used for these examples if they are compiled with the AMPL Solver Library [23].

- The CPLEX [11] solver, which is used to generate output for stochastic programming examples. This commercial solver provides capabilities needed for these examples that are not commonly available in open source optimization solvers (e.g., optimization of quadratic integer programs).
- The matplotlib Python package, which is used to generate plots.

Installation instructions for Pyomo are provided at the Pyomo website: www.pyomo.org. Appendix A provides a brief tutorial of the Python scripting language; various on-line sources provide more comprehensive tutorials and documentation.

1.4 Book Summary

The remainder of this book is divided into three parts. The first part provides an introduction to Pyomo. Chapter 2 provides a primer on optimization and mathematical modeling, including brief illustrations of how Pyomo can be used to specify and solve algebraic optimization models. Chapter 3 illustrates Pyomo's modeling capabilities with simple concrete and abstract models, and Chapter 4 describes Pyomo's core modeling components. The basics of embedding Pyomo models in scripts is in Chapter 5. The first part closes with Chapter 6 describing interaction with solvers.

The second part of this book documents advanced features and extensions. Chapter 7 describes the nonlinear programming capabilities of Pyomo, and Chapter 8 describes how hierarchical models can be expressed in Pyomo. Guidance on improving performance is given in Chapter 9. Chapter 10 describes the AbstractModel class, the syntax of Pyomo data command files, and Pyomo's command-line interface.

The third part of the book describes some of the modeling extensions. An overview of generalized disjunctive programming is provided in Chapter 11. Dynamic models expressed with differential and algebraic equations are described in Chapter 12, and programs with equilibrium constraints are described in Chapter 13).

> **NOTE:** This book does not provide a *complete* reference for Pyomo. Instead, our goal is to discuss core functionality that is available in the Pyomo 6.0 release.

1.5 Discussion

A variety of developers have realized that Python's clean syntax and rich set of supporting libraries make it an excellent choice for optimization modeling [30]. A variety of open source software packages provide optimization modeling capabilities in Python, such as PuLP [49], APLEpy [4], and OpenOpt [46]. Additionally, there are many Python-based solver interface packages, including open source packages such as PyGlpk [50] and pyipopt [51], in addition to Python interfaces for the commercial solvers such as CPLEX [11] and Gurobi [28].

Several features distinguish Pyomo. First, Pyomo provides mechanisms for extending the core modeling and optimization functionality without requiring edits to Pyomo itself. Second, Pyomo supports the definition of both concrete and abstract models. This allows the user significant flexibility in determining how closely data is integrated with a model definition. Finally, Pyomo can support a broad class of optimization models, including both standard linear programs as well as general nonlinear optimization models, generalized disjunctive programs, problems constrained by differential equations, and mathematical programs with equilibrium conditions.

Part I
An Introduction to Pyomo

Chapter 2
Mathematical Modeling and Optimization

Abstract This chapter provides a primer on optimization and mathematical modeling. It does not provide a complete description of these topics. Instead, this chapter provides enough background information to support reading the rest of the book. For more discussion of optimization modeling techniques see Williams [58]. Implementations of simple examples of models are shown to provide the reader with a quick start to using Pyomo.

2.1 Mathematical Modeling

2.1.1 Overview

Modeling is a fundamental process in many aspects of scientific research, engineering, and business. Modeling involves the formulation of a simplified representation of a system or real-world object. These simplifications allow structured representation of knowledge about the original system to facilitate the analysis of the resulting model. Schichl [56] notes models are used to:

- **Explain phenomena** arising in a system;
- **Make predictions** about future states of a system;
- **Assess key factors** influencing phenomena in a system;
- **Identify extreme states** in a system possibly representing worst-case scenarios or minimal cost plans; and
- **Analyze trade-offs** to support human decision makers.

Additionally, the structured aspect of a model's representation facilitates communication of the knowledge associated with a model. For example, a key aspect of a model is its level of detail, reflecting the system knowledge needed to employ the model in an application.

Mathematics has always played a fundamental role in representing and formulating our knowledge. Mathematical modeling has become increasingly formal as new

M. L. Bynum et al., *Pyomo — Optimization Modeling in Python*, Springer Optimization
and Its Applications 67, https://doi.org/10.1007/978-3-030-68928-5_2

frameworks have emerged to express complex systems. The following mathematical concepts are central to modern modeling activities:

- **Variables**: These represent *unknown* or changing parts of a model (e.g., decisions to take, or the characteristic of a system outcome).
- **Parameters**: These are symbolic representations for real-world data, and might vary for different problem instances or scenarios.
- **Relations**: These are *equations*, *inequalities*, or other mathematical relationships defining how different parts of a model are related to each other.

Optimization models are mathematical models with functions representing goals or objectives for the system being modeled. Optimization models can be analyzed to explore system trade-offs in order to find solutions to optimize system objectives. Consequently, these models can be used for a wide range of scientific, business, and engineering applications.

2.1.2 A Modeling Example

A *Model*, in the sense that we will use the word, represents items by abstracting away some features. Everyone is familiar with physical models, such as model railroads or model cars. Our interest is in mathematical models that use symbols to represent aspects of a system or real-world object.

For example, a person might want to determine the best number of scoops of ice cream to buy. We could use the symbol x to represent the number of scoops. We might use c to represent the cost per scoop. So then we could model the total cost as c times x, which we usually write as cx.

We might need a more sophisticated model of total cost if there are volume discounts or surcharges for buying fractional scoops. Also, this model is probably not valid for negative values of x. It is seldom possible to sell back ice cream for the same price paid for it.

It is more complicated to provide a mathematical model of the happiness associated with scoops of ice cream on an ice cream cone. One approach is to use a scaled measure of happiness. We will do that using the basic unit of the happiness associated with one scoop of ice cream, which we call h. A simple model, then, would be to say that the total happiness from x scoops of ice cream is h times x, which we write as hx. For some people, that might be a pretty good approximation for values of x between one-half and three, but there is almost no one who is 100 times as happy to have 100 scoops of ice cream on their ice cream cone as they are to have one scoop. For some people, the model of happiness for values of x between zero and ten might be something like

$$h \cdot \left(x - (x/5)^2\right).$$

Note that this model becomes negative when there are more than 25 scoops on the cone, which might not be a good model for everyone.

It is common to want to model more than one thing at a time. For example, you might be able to have scoops of ice cream and peanuts. Since there are multiple things that can be purchased, we can represent the quantity purchased using a vector x (i.e. the symbol x now represents a list). We refer to *elements* of the list using the notation x_i where the symbol i indexes the vector. For example, if we agree that the first element is the number of scoops of ice cream, then this number could be referenced using x_1. For higher dimensions a *tuple* is used, such as i, j or (i, j) as the index.

Let's change c to be a vector of costs with the same indices as x (i.e., c_1 is the cost per scoop of ice cream and c_2 is the cost per cup of peanuts). So now, we write the total cost of ice cream and peanuts as

$$c_1 x_1 + c_2 x_2 = \sum_{i=1}^{2} c_i x_i.$$

Once again, this cost model is probably not valid for all possible values of all elements of x, but it might be good enough for some purposes.

Often, it is useful to refer to indices as being members of a set. For the example just given, we could use the set $\{1, 2\}$ to write the total cost as

$$\sum_{i \in \{1,2\}} c_i x_i.$$

but it would be more common to use a more abstract expression like

$$\sum_{i \in \mathscr{A}} c_i x_i$$

where the set $\mathscr{A}$ is understood to be the index set for c and x (and for our example the set $\mathscr{A}$ would be $\{1, 2\}$.)

In addition to summing over an index set, we might want to have conditions that hold for all members of an index set. This is done simply by using a comma. For example, if we want to require that none of the values of x can be negative, we would write

$$x_i \geq 0, \; i \in \mathscr{A}$$

and we read this line out loud as "x subscript i is greater than or equal to zero for all i in A."

There is no law of mathematics or even mathematical modeling requiring the use of single letter symbols such as x and c or i. It would be perfectly okay for the set $\mathscr{A}$ to be composed of a picture of an ice cream cone and a picture of a cup of peanuts, but that is hard to work with in some settings. The set could also be $\{Scoops, Cups\}$, but that is not commonly done in books because it takes up too much space and causes lines to overflow. Also, x could be replaced by something like *Quantity*. Long names are, importantly, supported by modeling languages such as Pyomo, and it is generally a good idea to use meaningful names when writing Pyomo models. Spaces or dashes embedded in names often cause troubles and confusion, so underscores are often used in long names instead.

2.2 Optimization

The symbol x is often used as a *variable* in optimization modeling. It is sometimes called a *decision variable* because optimization models are built to help make decisions. This can sometimes cause a little confusion for people who are familiar with modeling as practiced by statisticians. They often use the symbol x to refer to data. Statisticians give values of x to the computer to have it compute statistics, while optimization modelers give other data to the computer and ask the computer to compute good values of x. Of course, symbols other than x can be used; though in textbooks and introductions x is often chosen.

Values such as cost (we used the symbol c) are referred to as *data* or *parameters*. An optimization model can be described with undefined parameter values, but a specific instance that is optimized must have specific data values, which we sometimes call *instance data*.

A model must have an objective to perform optimization, which is expressed as an *objective function*. Optimal values of the decision variables result in *the* best possible value of the objective function. It is important to note we did not say "the optimal values" because it is often the case that more than one set of variable values result in the best possible value of the objective function. It is common to write this function in a very abstract way, such as $f(x)$. Whether the best is the smallest or the largest possible value is determined by the *sense* of the optimization: *minimize* or *maximize*.

For example, suppose that x is not a vector, but rather a *scalar* denoting the number of scoops of ice cream to buy. If we use the model of happiness given before, then

$$f(x) \equiv h \cdot \left(x - (x/5)^2\right),$$

where h is given as data. (It turns out not to matter what value of h is given for the purpose of finding the x that maximizes happiness in this particular example.) The optimization problem, as we have modeled it, is given as

$$\max \ h \cdot \left(x - (x/5)^2\right),$$

but very careful authors would write

$$\max_{x} \ h \cdot \left(x - (x/5)^2\right)$$

to make it clear x is the decision variable. In this case, there is only one best value of x, which can be found using numerical optimization. The best value of x turns out to be fractional, which means that it is not an integer number of scoops. This model might not be considered useful for a typical ice cream shop, where the number of scoops must be a non-negative integer. To specify this requirement, we add a *constraint* to the optimization model:

$$\max_{x} h \cdot \left(x - (x/5)^2\right)$$

s.t.

$$x \in \text{non-negative integers}$$

where "s.t." is an abbreviation for either "subject to" or "such that." Suppose the model is not being used in an ice cream shop, but rather at home, where the ice cream is being served by the model user's parent. If the parent is willing to make partial scoops but not willing to go above two scoops, then the constraint

$$x \in \text{non-negative integers}$$

would be replaced with

$$0 \leq x \leq 2.$$

This is not a perfect model because really, not all fractional values of x would be reasonable.

To illustrate the model aspects discussed so far, let us return to multiple products described by an index set $\mathscr{A}$, so x is a vector. Let us make use of the following model of happiness for a product index i:

$$h_i \cdot \left(x_i - (x_i/d_i)^2\right),$$

where h and d are both data vectors with the same index set as x. Further, let c be a vector of costs and u be a vector of the most of any product that can be purchased. Assume that all products can be purchased in fractional quantities for the moment. Finally, suppose there is a total budget given by b. The optimization problem would be written as:

$$\max_{x} \sum_{i \in \mathscr{A}} h_i \cdot \left(x_i - (x_i/d_i)^2\right) \quad (H)$$

$$\text{s.t. } \sum_{i \in \mathscr{A}} c_i x_i \leq b$$

$$0 \leq x_i \leq u_i, \ i \in \mathscr{A}$$

Some modelers would express the last constraint separately:

$$\max_{x} \sum_{i \in \mathscr{A}} h_i \cdot \left(x_i - (x_i/d_i)^2\right) \quad (H)$$

$$\text{s.t. } \sum_{i \in \mathscr{A}} c_i x_i \leq b$$

$$x_i \leq u_i, i \in \mathscr{A}$$

$$x_i \geq 0, i \in \mathscr{A}$$

It is common to put a short, abbreviated name of the model in parentheses on the same line as the objective. The name (P) is very common, but we used (H) as a mnemonic for "happiness." The name (H) allows us to refer to this model later in the chapter, where we show how to implement it in Pyomo and solve it.

2.3 Modeling with Pyomo

We now consider different strategies for formulating and optimizing algebraic optimization models using Pyomo. Although a detailed explanation of Pyomo models is deferred to Chapter 3, the following examples illustrate the use of Pyomo for model (H).

2.3.1 A Concrete Formulation

A *concrete* Pyomo model initializes components as they are constructed. This allows modelers to easily make use of native Python data structures when defining a model instance. There are many ways to implement our model as a concrete Pyomo model, and we start with one using Python lists and dictionaries.

> **NOTE:** Recognizing that we will often make new instances of the model with different data, we choose to write a Python function that takes in the required data as arguments and returns a Pyomo model. Using this approach, enables us to reuse the general Pyomo model with different definitions of the data.

```python
import pyomo.environ as pyo

def IC_model(A, h, d, c, b, u):

    model = pyo.ConcreteModel(name = "(H)")

    def x_bounds(m, i):
        return (0,u[i])
    model.x = pyo.Var(A, bounds=x_bounds)

    def z_rule(model):
        return sum(h[i] * (model.x[i] - (model.x[i]/d[i])**2)
                for i in A)
    model.z = pyo.Objective(rule=z_rule, sense=pyo.maximize)

    model.budgetconstr = pyo.Constraint(\
            expr = sum(c[i]*model.x[i] for i in A) <= b)

    return model
```

In the budgetconstr declaration, we define the constraint directly with the expr keyword argument, however, a construction rule could also be used. An example of a construction rule is shown in the declaration of the objective function. We could have used an expr keyword in the objective function declaration.

> **NOTE:** The backslash character at the end of a line tells Python that the line continues; we use it to help make the lines fit on a book page. In this particular case it is not strictly required because the line is breaking inside a parenthetical grouping.

There are more elegant ways to create this IC_Model function, but the function as written provides an illustration of some of the choices that can be made. With particular data in hand, one can write a Python program that provides the data to the function to obtain the fully instantiated Pyomo model. If there is a solver installed on the computer, the Python program can then send the model to a solver to be solved and, if successful, query the model for the solution. But before we look into these steps, let's consider more of the many ways this function could be written.

Note the function IC_model is just a Python function. One could have written the Pyomo model directly in the Python program that calls the solver, or defined a function IC_model to accept a dictionary as an argument instead of the more explicit argument list. Python programmers can probably think of more, and better, ways to write this Python code.

In addition to the modeling components already discussed, Pyomo also offers sets (Set) and parameters (Param), which are components that will be discussed in subsequent chapters. In the code below, we have defined the function IC_model_dict that takes in a Python dictionary and makes use of Set and Param objects to define the same model.

```python
import pyomo.environ as pyo

def IC_model_dict(ICD):
    # ICD is a dictionary with the data for the problem

    model = pyo.ConcreteModel(name = "(H)")

    model.A = pyo.Set(initialize=ICD["A"])

    model.h = pyo.Param(model.A, initialize=ICD["h"])
    model.d = pyo.Param(model.A, initialize=ICD["d"])
    model.c = pyo.Param(model.A, initialize=ICD["c"])
    model.b = pyo.Param(initialize=ICD["b"])
    model.u = pyo.Param(model.A, initialize=ICD["u"])

    def xbounds_rule(model, i):
        return (0, model.u[i])
    model.x = pyo.Var(model.A, bounds=xbounds_rule)

    def obj_rule(model):
        return sum(model.h[i] * \
```

```
            (model.x[i] - (model.x[i]/model.d[i])**2)\
                for i in model.A)
    model.z = pyo.Objective(rule=obj_rule,sense=pyo.maximize)

    def budget_rule(model):
        return sum(model.c[i]*model.x[i]\
                for i in model.A) <= model.b
    model.budgetconstr = pyo.Constraint(rule=budget_rule)

    return model
```

2.4 Linear and Nonlinear Optimization Models

2.4.1 Definition

An expression in an optimization model is said to be linear if it is composed only of sums of decision variables and/or decision variables multiplied by data. Accordingly, a linear expression is a non-constant, linear function of the decision variables. Assume x is a variable vector, c is a vector of data and are indexed by $\mathscr{A}$. Further assume 2 and 3 are members of $\mathscr{A}$. The following are linear expressions:

$$\sum_{i \in \mathscr{A}} c_i x_i$$
$$\sum_{i \in \mathscr{A}} x_i$$
$$x_2$$
$$c_3 x_2 + c_2 x_3$$
$$c_3 x_2 + c_2 x_3 + 4$$

On the other hand, the following expressions are not linear: x_i^2, $x_2 x_3$ and $\cos(x_2)$.

Linear expressions often result in problems that can be solved with much less computational effort than similar models with nonlinear expressions. Consequently, many modelers make an effort to use linear expressions as much as possible, and some modelers strive to use only linear expressions. Additionally, many modelers develop linear approximations to nonlinear models in hopes of finding "good enough" solutions to the original nonlinear model.

For illustrative purposes, let's assume we have the following linear approximation to (H), and we will replace the objective function in (H) with

$$\max_x \sum_{i \in \mathscr{A}} h_i \cdot \left(1 - u_i/d_i^2\right) x_i, \tag{2.1}$$

where u_i is a new model parameter. We say that this expression is linear because the decision variables are only multiplied by data, and summed. It is true the parameter d is squared, but this is not a decision variable. The numerical value of the entire expression

$$h_i \cdot \left(1 - u_i/d_i^2\right)$$

is computed by Pyomo before the problem instance is passed to a solver, and the task of the solver is to find optimal values for the decision variables.

2.4.2 Linear Version

If we want to modify the concrete model given on page 20 to use expression (2.1), we would change the objective function expression rule as follows:

```
def obj_rule(model):
    return sum(h[i]*(1 - u[i]/d[i]**2) * model.x[i] \
            for i in A)
```

2.5 Solving the Pyomo Model

Pyomo provides automated methods to (1) combine the model and data, (2) send the resulting *model instance* to a solver, and (3) recover the results for display and further use. Pyomo does not, itself, solve optimization problem instances. They are always passed to a solver of some sort.

2.5.1 Solvers

Pyomo can be installed without any solvers. For example, Pyomo can simply write out problem instances into files suitable as direct input to a solver. This use of Pyomo might be necessary if the solver is run separately on a different computer. Typically, a solver should be installed and accessible to Pyomo, and most of the examples in this book make this assumption.

Recall the objective in (H) is not a linear function of the variable, x, and the budget constraint is linear. Although many solvers can solve an instance with a quadratic objective and linear constraints, some solvers cannot. If the only solver on the computer is limited to linear problems, then (H) would need to approximate with a linear model.

2.5.2 Python Scripts

A Python script is executed using Python from the command line or within a development environment. As with expressing the model, there are many options for writing a script to supply its data and to solve it described in subsequent chapters. For example, a script defining the concrete model given on page 2.3.1 can be created by adding the following lines:

```
A = ['I_C_Scoops', 'Peanuts']
h = {'I_C_Scoops': 1, 'Peanuts': 0.1}
d = {'I_C_Scoops': 5, 'Peanuts': 27}
c = {'I_C_Scoops': 3.14, 'Peanuts': 0.2718}
b = 12
u = {'I_C_Scoops': 100, 'Peanuts': 40.6}

model = IC_model_linear(A, h, d, c, b, u)

opt = pyo.SolverFactory('glpk')
results = opt.solve(model) # solves and updates model
pyo.assert_optimal_termination(results)

model.display()
```

If the resulting file is called `ConcHLinScript.py`, then it can be run from the terminal with the command line:

```
python ConcHLinScript.py
```

The first few lines that assign data to Python variables look a little strange, because they are. Usually, data for optimization is read from files or databases; however, for this textbook example we assign the data using Python literals so it is self-contained. The last few lines create a solver, solve the model, and display the model with the solution values. The function `assert_optimal_termination` halts the script and outputs a message if the solver does not report it found an optimal solution. The companion function `check_optimal_termination` returns `True` if the solver reports optimality and `False` if not.

Chapter 3
Pyomo Overview

Abstract This chapter provides an overview of the modeling strategies and capabilities of Pyomo. A brief discussion of the core modeling components supported by Pyomo, and some of the modeling capabilities within Pyomo (e.g., discrete variables and nonlinear models) are provided.

3.1 Introduction

Pyomo supports an object-oriented design for the definition of optimization models. A Pyomo *model* object contains a collection of modeling *components* defining the optimization problem. The Pyomo package includes modeling components necessary to formulate an optimization problem: variables, objectives, and constraints, as well as other modeling components commonly supported by modern AMLs, including index sets and parameters. These basic modeling components are defined in Pyomo through the following Python classes:

`Var`	optimization variables in a model
`Objective`	expressions that are minimized or maximized in a model
`Constraint`	constraint expressions in a model
`Set`	set data that is used to define a model instance
`Param`	parameter data that is used to define a model instance

In this chapter, an overview of these components and the process to define and solve Pyomo models is given. The basic steps of a simple modeling process are as follows:

1. Create an instance of a model using Pyomo modeling components.
2. Pass this instance to a solver to find a solution.
3. Report and analyze results from the solver.

Pyomo supports general scripting with Python where a user can flexibly control the solution process and develop a custom workflow, such as solving sequences of problems with modifications, or more complex meta-algorithms.

M. L. Bynum et al., *Pyomo — Optimization Modeling in Python*, Springer Optimization and Its Applications 67, https://doi.org/10.1007/978-3-030-68928-5_3

In this chapter, an example problem is used to illustrate the process of formulating a real-world model, including the use of modeling components, indexed components, and construction rules. The use of scripting for more advanced workflows is also discussed.

3.2 The Warehouse Location Problem

The warehouse location problem is used throughout this chapter. This formulation seeks to find the locations for a set of warehouses that meet delivery demands while optimizing transportation costs. Let N be a set of candidate warehouse locations, and let M be a set of customer locations. For each warehouse n, the cost of delivering product to customer m is given by $d_{n,m}$. The goal is to determine the optimal warehouse locations that will minimize the total cost of product delivery. The binary variables y_n are used to define whether or not a warehouse should be built, where y_n is 1 if warehouse n is selected and 0 otherwise. The variable $x_{n,m}$ indicates the fraction of demand for customer m that is served by warehouse n.

The variables x and y are to be determined by the optimization solver, while all other quantities are known inputs or parameters in the problem. This problem is a particular description of the *p-median* problem, and it has the interesting property that there will be optimal x values in $\{0,1\}$ even though they are not specified as binary variables.

The complete problem formulation is:

$$\min_{x,y} \sum_{n\in N}\sum_{m\in M} d_{n,m}x_{n,m} \tag{WL.1}$$

$$\text{s.t.} \sum_{n\in N} x_{n,m} = 1, \ \forall\, m \in M \tag{WL.2}$$

$$x_{n,m} \le y_n, \ \forall\, n \in N, \, m \in M \tag{WL.3}$$

$$\sum_{n\in N} y_n \le P \tag{WL.4}$$

$$0 \le x \le 1 \tag{WL.5}$$

$$y \in \{0,1\} \tag{WL.6}$$

Here, the objective (equation WL.1) is to minimize the total cost associated with delivering products to all the customers. Equation WL.2 ensures that each customer's demand is fully met, and equation WL.3 ensures that a warehouse can deliver product to customers only if that warehouse is selected to be built. With equation WL.4 the number of warehouses that can be built is limited to P.

For our example, we will assume that $P=2$, with the following data for warehouse and customer locations,

Customer locations $\quad\quad\quad\quad\quad$ $M = \{$'NYC', 'LA', 'Chicago', 'Houston'$\}$
Candidate warehouse locations $\quad$ $N = \{$'Harlingen', 'Memphis', 'Ashland'$\}$

with the costs $d_{n,m}$ as given in the following table:

	NYC	LA	Chicago	Houston
Harlingen	1956	1606	1410	330
Memphis	1096	1792	531	567
Ashland	485	2322	324	1236

3.3 Pyomo Models

Pyomo supports an object-oriented design where modeling components are added to a Pyomo model to define the optimization problem. In this section, an overview of the common modeling components is given, and complete Pyomo examples of the warehouse location problem are provided.

3.3.1 Components for Variables, Objectives, and Constraints

Optimization problems require, at least, one variable and an objective function. Most problems also include constraints. The Pyomo classes for implementing these modeling components are `Var`, `Objective`, and `Constraint`. The following example shows how these components could be defined:

```
model.x = pyo.Var()
model.y = pyo.Var(bounds=(-2,4))
model.z = pyo.Var(initialize=1.0, within=pyo.NonNegativeReals)

model.obj = pyo.Objective(expr=model.x**2 + model.y + model.z)

model.eq_con = pyo.Constraint(expr=model.x + model.y + model.z \
    == 1)
model.ineq_con = pyo.Constraint(expr=model.x + model.y <= 0)
```

This example includes three optimization variables (x, y, and z), a single objective, and two constraints. For each optimization variable, an instance of the `Var` class is created and that instance is added as an attribute to the model object. The code `model.x=pyo.Var()` creates an instance of the Pyomo class `Var` and assigns it to `model.x`. The `model` object identifies when a component is being added and performs special processing that includes, for example, setting the name of the instance of `Var` to "x", and setting a reference to the owning model.

This example declares x as a continuous variable, but keyword arguments can be used to define different properties of the Pyomo `Var`. For example, bounds is used to set lower and upper bounds, `initialize` is used to set initial values, and `within` is used to set the domain. In this example, `model.y` has a lower bound of -2 and an upper bound of 4, and `model.z` has a lower bound of 0, and no upper bound since the keyword argument `within` is set to non-negative reals.

> **NOTE:** The use of keyword arguments is common in the constructors for Pyomo components to specify component properties. See Chapter 4 for more details about supported keyword arguments for Pyomo components.

This example defines an objective function using the `Objective` component. The `expr` keyword is used to define the expression for the objective function. By default, optimization objectives are minimized, but the `sense` keyword can be set to `maximize` for maximization problems. This example declares an equality constraint and inequality constraint using `Constraint` components. The `expr` keyword is used again to define the mathematical expressions for the constraints, including the logical operator separating the left hand side expression and the right hand side expression. Constraints can include a logical operator to equal to (==), less than or equal to (<=), or greater than or equal to (>=). See Chapter 4 for a detailed description of the `Objective` and `Constraint` components and the available keyword arguments.

> **NOTE:** In the previous example, the objective and constraints were defined with the `expr` keyword. While this is convenient for illustrating the examples with few lines of code, these components are often defined using *construction rules*, which are discussed in more detail in Sections 3.3.3 and 4.2.1.

3.3.2 Indexed Components

In the previous example, each of the modeling components was *scalar*. Specifically, each of the optimization variables x, y, and z were single values only, not vectors or arrays. The constraints were also scalar, where each declaration created only a single mathematical constraint. When modeling large, complex applications, it is common to have vectors of variables and constraints whose dimension and indexing is determined according to model data. This is handled within Pyomo through *indexed* components.

To illustrate the concept of an indexed component, consider the warehouse location problem (WL) defined using only scalar components. Note, a better approach using indexed components will be shown subsequently. For example, separate x variables could be created for each pair of warehouses and customers,

```
model.x_Harlingen_NYC = pyo.Var(bounds=(0,1))
model.x_Harlingen_LA = pyo.Var(bounds=(0,1))
model.x_Harlingen_Chicago = pyo.Var(bounds=(0,1))
model.x_Harlingen_Houston = pyo.Var(bounds=(0,1))
model.x_Memphis_NYC = pyo.Var(bounds=(0,1))
model.x_Memphis_LA = pyo.Var(bounds=(0,1))
#...
```

and, the constraint described in WL.4 manually expanded as,

```
model.maxY = pyo.Constraint(expr=model.y_Harlingen + \
    model.y_Memphis + model.y_Ashland <= P)
```

and, all the constraints in equation (WL.2) could them be explicitly written as,

```
model.one_warehouse_for_NYC = \
    pyo.Constraint(expr=model.x_Harlingen_NYC + \
    model.x_Memphis_NYC + model.x_Ashland_NYC == 1)

model.one_warehouse_for_LA = \
    pyo.Constraint(expr=model.x_Harlingen_LA + \
    model.x_Memphis_LA + model.x_Ashland_LA == 1)
#...
```

However, this would become very cumbersome for large data sets, and this is much easier to formulate using *indexed* components. First, we can define a list of valid indices for the warehouse locations and the customer locations,

```
N = ['Harlingen', 'Memphis', 'Ashland']
M = ['NYC', 'LA', 'Chicago', 'Houston']
```

and, using this data, we can then declare variables as follows:

```
model.x = pyo.Var(N, M, bounds=(0,1))
model.y = pyo.Var(N, within=pyo.Binary)
```

We refer to N and M as index sets for the indexed variables model.x and model.y. Specifically, the variable y is indexed over N, and the variable x is a two-dimensional array that is indexed over both N and M. With this declaration, an element of x can be accessed by model.x[i,j] where i and j are elements of the sets N and M, respectively.

> **NOTE:** Pyomo modeling components can include any number of index sets as unnamed arguments in their declaration but they must be specified before any other named keyword arguments. These index sets specify the valid indices for individual elements of the component.

Given these declarations, constraint (WL.4) can be defined as

```
model.num_warehouses = pyo.Constraint(expr=sum(model.y[n] for n \
    in N) <= P)
```

This declaration uses Python's iteration syntax to sum over a set of indexed variables. The *list comprehension* syntax enables a concise specification of the summation, where the syntax specifies that the terms `model.y[n]` are generated by iterating over the set N. As these terms are generated, the function `sum` adds them together to form the overall expression. Similarly, the objective can be defined as

```
model.obj = pyo.Objective(expr=sum(d[n,m]*model.x[n,m] for n in \
    N for m in M))
```

where the terms `d[n,m]*model.x[n,m]` are generated by iterating over both N and M.

3.3.3 Construction Rules

The construction of many indexed constraints is performed with *construction rules*. Consider constraint (WL.2):

$$\sum_{n \in N} x_{n,m} = 1, \ \forall \, m \in M$$

This mathematical notation indicates there is a single constraint defined for each m in the set M. The `Constraint` component can be declared as an indexed constraint over the elements in this set. However, a mechanism is needed to provide Pyomo with the explicit expressions for each element in M. Pyomo allows model components to be initialized with user-defined functions called *rules*.

The following example illustrates the use of a construction rule to define constraint (WL.2):

```
def demand_rule(mdl, m):
    return sum(mdl.x[n,m] for n in N) == 1
model.demand = pyo.Constraint(M, rule=demand_rule)
```

The first two lines define a Python function that will be called to produce the correct constraint expression for each element in M. The last line in this example declares the constraint by creating a `Constraint` component that is indexed over the set defined by M. The `rule` keyword argument indicates that the function `demand_rule` will be called to construct each constraint.

The first argument in the function `demand_rule` will automatically be set to the instance of the model object being constructed. It is followed by arguments providing the indices of the particular constraint being constructed. When Pyomo constructs the `Constraint` object, the construction rule is called once for each of the values of the specified index sets.

NOTE: Pyomo expects a construction rule to return an expression for every index value. If no constraint is needed for a particular combination of indices, then the value `Constraint.Skip` can be returned instead.

Construction rules can be used for most modeling components, using the `rule` keyword argument, even if the component is not indexed. Although the function arguments for component rules are the same for all component types, the following table illustrates that the expected type of the return value is different:

Component	Construction Rule Return Types
Set	A Python set or list object
Param	An integer or float value
Objective	An expression
Constraint	A constraint expression.

3.3.4 A Concrete Model for the Warehouse Location Problem

The warehouse location problem can be defined as a concrete model as follows:

```python
1   # wl_concrete.py
2   # ConcreteModel version of warehouse location problem
3   import pyomo.environ as pyo
4
5   def create_warehouse_model(N, M, d, P):
6       model = pyo.ConcreteModel(name="(WL)")
7
8       model.x = pyo.Var(N, M, bounds=(0,1))
9       model.y = pyo.Var(N, within=pyo.Binary)
10
11      def obj_rule(mdl):
12          return sum(d[n,m]*mdl.x[n,m] for n in N for m in M)
13      model.obj = pyo.Objective(rule=obj_rule)
14
15      def demand_rule(mdl, m):
16          return sum(mdl.x[n,m] for n in N) == 1
17      model.demand = pyo.Constraint(M, rule=demand_rule)
18
19      def warehouse_active_rule(mdl, n, m):
20          return mdl.x[n,m] <= mdl.y[n]
21      model.warehouse_active = pyo.Constraint(N, M, \
                rule=warehouse_active_rule)
22
23      def num_warehouses_rule(mdl):
24          return sum(mdl.y[n] for n in N) <= P
25      model.num_warehouses = \
                pyo.Constraint(rule=num_warehouses_rule)
26
27      return model
```

This file begins by importing the Pyomo environment, which defines the Python classes used to build a model. Line 5 defines a function that will be called to create and return the model. This is not necessary, and the model can be created directly in a Python script, however, this strategy is often preferred so this model construction code can be easily reused with different data. Line 6 creates the `ConcreteModel` and provides a name.

Lines 8 and 9 declare and construct the variables for the problem. The `model` object is a `ConcreteModel`, and once these lines are executed, the variables x and y are completely constructed with known indices. Lines 11 and 12 define the *construction rule* for the objective function, and line 13 declares the objective function and assigns it to `model.obj`. As soon as line 13 executes, the rule declared on lines 11 and 12 is called to construct the expression for the objective function. Similarly, in the remaining lines of the Python file, the constraint rules are declared, followed by the constraint objects themselves. Since this is a `ConcreteModel`, the constraint rules are called when Python executes lines 17, 21, and 25. Line 27 returns the constructed model from the function.

Now that the model is defined, we can create a short Python script that solves a particular instance of the model and shows the solution.

```python
1    # wl_concrete_script.py
2    # Solve an instance of the warehouse location problem
3
4    # Import Pyomo environment and model
5    import pyomo.environ as pyo
6    from wl_concrete import create_warehouse_model
7
8    # Establish the data for this model (this could also be
9    # imported using other Python packages)
10
11   N = ['Harlingen', 'Memphis', 'Ashland']
12   M = ['NYC', 'LA', 'Chicago', 'Houston']
13
14   d = {('Harlingen', 'NYC'): 1956, \
15        ('Harlingen', 'LA'): 1606, \
16        ('Harlingen', 'Chicago'): 1410, \
17        ('Harlingen', 'Houston'): 330, \
18        ('Memphis', 'NYC'): 1096, \
19        ('Memphis', 'LA'): 1792, \
20        ('Memphis', 'Chicago'): 531, \
21        ('Memphis', 'Houston'): 567, \
22        ('Ashland', 'NYC'): 485, \
23        ('Ashland', 'LA'): 2322, \
24        ('Ashland', 'Chicago'): 324, \
25        ('Ashland', 'Houston'): 1236 }
26   P = 2
27
28   # Create the Pyomo model
29   model = create_warehouse_model(N, M, d, P)
30
31   # Create the solver interface and solve the model
32   solver = pyo.SolverFactory('glpk')
33   res = solver.solve(model)
```

```
34   pyo.assert_optimal_termination(res)
35
36   model.y.pprint() # Print the optimal warehouse locations
```

Line 5 imports the Pyomo environment, and line 6 imports the function defined in wl_concrete.py to create the model from the passed in data. Lines 11 through 26 define the data for this problem. The Python lists N and M are used to specify the valid warehouse locations and the customer locations respectively. The Python dictionary d defines the costs associated with serving each customer from each location, and line 26 specifies P, providing the number of warehouses needed.

In line 29 these native Python data structures are passed to the function written, create_warehouse_location, where they are used to declare and construct the Pyomo modeling components, the Var, Objective, and Constraint objects. The constructed model is returned from the function and assigned to model. Line 32 creates an interface to the solver "glpk" that can be used to solve the optimization problem. Line 33 calls solve to execute the solver and return a results object to res, which is passed to the function assert_optimal_termination in line 34. If the solver does not report an optimal solution (possibly because the solver is not properly installed or because there is not an optimal solution), this function will print a message and terminate the script.

> **NOTE:** After a Pyomo model has been constructed, the model can be printed using the pprint method, model.pprint(). This summarizes the information in the Pyomo model, including the constraint and objective expressions. This can be a very useful debugging tool when a model is not generating the expected results, since it shows the fully expanded version of the model.

In this example, the Python data for the problem (N, M, d, and P) were explicitly defined in the script. While this is convenient to create a short book example, in practice, much more data is often required, and this data would instead be loaded from another source (e.g., an Excel file, or JSON file).

Consider Figure 3.1, showing some example data for the warehouse location problem specified in Microsoft Excel. The following script loads this data from the Excel spreadsheet using the Python package Pandas, and then executes the same lines as before to construct and solve the model, and report the solution.

```python
# wl_excel.py: Loading Excel data using Pandas
import pandas
import pyomo.environ as pyo
from wl_concrete import create_warehouse_model

# read the data from Excel using Pandas
df = pandas.read_excel('wl_data.xlsx', 'Delivery Costs', \
    header=0, index_col=0)

N = list(df.index.map(str))
```

```
M = list(df.columns.map(str))
d = {(r, c):df.at[r,c] for r in N for c in M}
P = 2

# create the Pyomo model
model = create_warehouse_model(N, M, d, P)

# create the solver interface and solve the model
solver = pyo.SolverFactory('glpk')
solver.solve(model)

model.y.pprint() # print the optimal warehouse locations
```

	A	B	C	D	E	F
1		NYC	LA	Chicago	Houston	
2	Harlingen	1956	1606	1410	330	
3	Memphis	1096	1792	531	567	
4	Ashland	485	2322	324	1236	
5						
6						

Fig. 3.1: This figure shows the data for our warehouse location problem as formatted in Microsoft Excel.

3.3.5 Modeling Components for Sets and Parameters

While data can be specified using native Python types, Pyomo also includes modeling components Set and Param to define index sets and parameters respectively.

A Pyomo Set component is used to declare valid indices for any indexed component. For example, in the context of the warehouse location problem, two sets are shown: N stores the valid warehouse locations and M stores the customer locations. These sets can easily be declared using the following code:

```
model.N = pyo.Set()
model.M = pyo.Set()
```

These Set objects can be used to define indexed variables or constraints:

```
model.x = pyo.Var(model.N, model.M, bounds=(0,1))
model.y = pyo.Var(model.N, within=pyo.Binary)
```

This example passes Set objects into the Var constructor, rather than the Python lists used in earlier examples. A Set component can be initialized by employing the initialize keyword argument, with a Python set, list, or tuple.

Pyomo `Set` objects can also be indexed by other sets. Consider the following example:

```
model.PremierSundaes = pyo.Set()
model.Toppings = pyo.Set(model.PremierSundaes)
```

The set `model.Toppings` is an indexed set. If `model.PremierSundaes` is given the values { 'PBC-Banana', 'Very Berry' }, then toppings for each of these indices can be defined. For example, `model.Toppings['PBC-Banana']` might contain the set { 'Peanut Butter', 'Chocolate Fudge', 'Banana' }. On the other hand, `model.Toppings['Very Berry']` might contain {'Strawberries', 'Raspberries', 'Blueberries', 'Crunch-berries'}.

A Pyomo `Param` component can be used to define data values for this problem. In the context of the warehouse location problem, two pieces of data need to be specified: P and $d_{n,m}$. These parameters can be declared using the following code:

```
model.d = pyo.Param(model.N, model.M)
model.P = pyo.Param()
```

This example declares a scalar parameter P and an indexed parameter d. The parameter d is indexed by the Pyomo sets for the warehouse and customer locations defined earlier. As with the `Set` object, values for these parameters could be provided through the `initialize` keyword argument using a Python dictionary or by defining a construction rule.

By default, parameters are immutable, meaning once their values are set, these values cannot be changed. This default behavior allows for increased efficiency within Pyomo when handling expressions. However, a parameter whose values are mutable can be defined with the `mutable=True` keyword argument. This can be useful if a model should be solved multiple times with different values of some of the parameters.

As an example, consider the warehouse location problem again. Assume significantly more data is required (e.g, a large number of potential warehouse locations and customer locations). Using a mutable parameter for P easily shows how the optimal delivery costs change when the maximum number of warehouses is changed.

The following code shows the process to define the model with a mutable parameter for P using the Pyomo `Param` object.

```
# wl_mutable.py: warehouse location problem with mutable param
import pyomo.environ as pyo

def create_warehouse_model(N, M, d, P):
    model = pyo.ConcreteModel(name="(WL)")

    model.x = pyo.Var(N, M, bounds=(0,1))
    model.y = pyo.Var(N, within=pyo.Binary)
    model.P = pyo.Param(initialize=P, mutable=True)

    def obj_rule(mdl):
        return sum(d[n,m]*mdl.x[n,m] for n in N for m in M)
    model.obj = pyo.Objective(rule=obj_rule)
```

```
    def demand_rule(mdl, m):
        return sum(mdl.x[n,m] for n in N) == 1
    model.demand = pyo.Constraint(M, rule=demand_rule)

    def warehouse_active_rule(mdl, n, m):
        return mdl.x[n,m] <= mdl.y[n]
    model.warehouse_active = pyo.Constraint(N, M, \
        rule=warehouse_active_rule)

    def num_warehouses_rule(mdl):
        return sum(mdl.y[n] for n in N) <= mdl.P
    model.num_warehouses = \
        pyo.Constraint(rule=num_warehouses_rule)

    return model
```

The key differences are the declaration of the `Param` object, `model.P`, and the use of `model.P` in the num_warehouses constraint. The script can be modified to load the distance data from Excel, and execute a loop in Python to solve the optimization problem repeatedly for different values of the mutable parameter `model.P`. This script is shown below.

```
# wl_mutable_excel.py: solve problem with different values for P
import pandas
import pyomo.environ as pyo
from wl_mutable import create_warehouse_model

# read the data from Excel using Pandas
df = pandas.read_excel('wl_data.xlsx', 'Delivery Costs', \
    header=0, index_col=0)

N = list(df.index.map(str))
M = list(df.columns.map(str))
d = {(r, c):df.at[r,c] for r in N for c in M}
P = 2

# create the Pyomo model
model = create_warehouse_model(N, M, d, P)

# create the solver interface
solver = pyo.SolverFactory('glpk')

# loop over values for mutable parameter P
for n in range(1,10):
    model.P = n
    res = solver.solve(model)
    pyo.assert_optimal_termination(res)
    print('# warehouses:', n, \
        'delivery cost:', pyo.value(model.obj))
```

Pyomo users can leverage Python's powerful scripting capabilities to execute custom workflows that manipulate and optimize models. This section has only scratched the surface of the possibilities with scripting. More details are provided in Chapter 5.

Chapter 4
Pyomo Models and Components: An Introduction

Abstract This chapter describes the core classes used to define optimization models in Pyomo. Most of the discussion focuses on modeling components used to declare parts of a model. Included is a discussion of the options used when declaring the components and information about key component attributes and methods.

4.1 An Object-Oriented AML

Pyomo supports an object-oriented approach for representing mathematical optimization models. A model object is created, and then modeling components are added to this object to declare different parts of the model. Pyomo includes modeling components commonly supported by modern AMLs: variables, constraints, objectives, index sets, and symbolic parameters. This chapter describes the Pyomo modeling components. In subsequent chapters, additional components are introduced that provide enhanced functionality to represent advanced optimization model features.

Users can create two types of models in Pyomo: concrete and abstract. A *concrete* model is constructed "on-the-fly" as each model component is declared. Therefore, the data associated with a concrete model must be specified before model components are declared. A user can leverage native Python data structures to define components in a concrete model. The `ConcreteModel` class is used to represent a concrete model.

In contrast, an *abstract* model supports complete declaration of a model abstractly. A specific problem instance is not constructed until all components are declared and the data is provided. The `AbstractModel` class is used to create an abstract model. Because abstract models allow components to reference data before it is defined, they often rely on Pyomo data components such as `Set` and `Param` to provide an abstract definition of the data used to construct the model (although these components can be used on concrete models as well).

M. L. Bynum et al., *Pyomo — Optimization Modeling in Python*, Springer Optimization
and Its Applications 67, https://doi.org/10.1007/978-3-030-68928-5_4

37

The following are the core modeling components in Pyomo:

Var   The Var component is used to represent optimization variables. Py-
               omo supports continuous and discrete variables, and includes sev-
               eral pre-defined domains.

Objective   The Objective component defines the function or functions to
               be optimized by the solver. This component contains the expression
               used to define the objective function, and a flag to indicate the sense
               (maximize or minimize).

Constraint   Constraints are used to define additional restrictions on the op-
               timization variables. The Constraint component contains ex-
               pressions and the appropriate relational operator. Pyomo supports
               equality (==) and general inequality (<= or >=) constraints.

Set   The Set component represents a collection of data that can include
               numeric (e.g., integer), or symbolic (e.g., string) elements. They are
               most commonly used to define valid indices for other components.
               Several common set operations are also supported.

Param   The Param component is used to represent numerical or symbolic
               values for data in the optimization problem. In contrast with simple
               Python data types (e.g., float), Param objects support the ability
               to change values (meaning they are *mutable*), and include features
               like sparse representations and default values.

Expression   The Expression component can be used to create a Pyomo ex-
               pression that can be reused in different parts of a Pyomo model.
               This is useful for representing common sub-expressions for mem-
               ory efficiency. Similar to mutable parameters, the underlying ex-
               pression can be changed between calls to the solver.

Suffix   Frequently, there is a need to provide, or receive, meta-data about
               a model or a component (e.g., dual information from a constraint).
               This is supported through the Pyomo Suffix component.

This chapter describes each of these components in more detail. A variety of other
modeling components are included in Pyomo, some of which are briefly discussed
at the end of this chapter and covered in more detail in the remaining chapters of
this book.

> **NOTE:** Unless otherwise stated, the code snippets and examples used in this
> chapter refer to concrete models.

4.2 Common Component Paradigms

There are behaviors common across most of the Pyomo modeling components listed
in the previous section. Additionally, there are some common paradigms adopted

across many components. This section describes these common behaviors.

4.2.1 Indexed Components

As shown in the previous chapter, Pyomo components can be declared as individual, atomic entities or as indexed collections. Indexed components will appear in several of the examples in this chapter. Consider the following model:

```
model = pyo.ConcreteModel()
model.A = pyo.Set(initialize=[1,2,3])
model.B = pyo.Set(initialize=['Q', 'R'])
model.x = pyo.Var()
model.y = pyo.Var(model.A, model.B)
model.o = pyo.Objective(expr=model.x)
model.c = pyo.Constraint(expr=model.x >= 0)
def d_rule(model, a):
    return a * model.x <= 0
model.d = pyo.Constraint(model.A, rule=d_rule)
```

The component c specifies a single constraint in this model, and the component d specifies a collection of constraints indexed over the set A. The Constraint component can be used to declare both simple constraints and indexed constraints. In general, components can also be indexed by multiple index sets. For example, model.y is indexed over both A and B, and it can be referenced by model.y[i,j] where i is any valid element from model.A and j is any valid element from model.B (e.g., model.y[2,'Q']).

> **NOTE:** Any unnamed arguments in a component constructor are assumed to be index sets for the component. They specify the set of valid indices for the component.

Declaration of arguments for indexed components is often more complex. For example, the initialize keyword argument can be used when declaring a single variable,

```
model.x = pyo.Var(initialize=3.14)
```

Specifying a value for these types of keyword arguments is straightforward when the component is not indexed. When the component is indexed, however, we may want to specify a different value for each of the indices. There are three approaches typically supported for these kinds of keyword arguments.

- When a single scalar value is passed, then that value is used for all the indices of the component.
- In many cases, you can also pass a Python dictionary (index-value pairs) where the keys of the dictionary are valid indices for the component.

- It is also possible to pass in a Python function to provide the value for every index in the component. We often call these functions rules.

These uses are illustrated here:

```
model.A = pyo.Set(initialize=[1,2,3])
model.x = pyo.Var(model.A, initialize=3.14)
model.y = pyo.Var(model.A, initialize={1:1.5, 2:4.5, 3:5.5})
def z_init_rule(m, i):
    return float(i) + 0.5
model.z = pyo.Var(model.A, initialize=z_init_rule)
```

4.3 Variables

Pyomo variables are created using the `Var` class, which can represent a single value or an indexed collection of values. Variables can have initial values, and the value of a variable can be retrieved and set by the user or by a solver as part of the solution process.

4.3.1 *Var Declarations*

The following code creates a non-indexed `Var` object:

```
model.x = pyo.Var()
```

Named and un-named arguments are supported, and Table 4.1 provides a list of the common arguments that can be passed when declaring the `Var` component

keyword	description	acceptable values
\<un named\>	reserved for specifying index sets	any number of Pyomo `Set` objects or Python lists
`within` or `domain`	specifies the valid domain or values for a variable	a Pyomo `Set` object, Python list, or rule function
`bounds`	provides lower and upper bounds for the variable	a 2-tuple, or a rule function
`initialize`	provides an initial value for the variable	a scalar value, Python dictionary of index-value pairs, or rule function

Table 4.1: Common declaration arguments for the `Var` component

The domain of a variable (i.e., the set of legal values) is specified with either the `domain` or `within` keyword options to the `Var` constructor:

```
model.A = pyo.Set(initialize=[1,2,3])
model.y = pyo.Var(within=model.A)
model.r = pyo.Var(domain=pyo.Reals)
model.w = pyo.Var(within=pyo.Boolean)
```

In this example, `y` is only allowed to take on the integer values 1, 2, or 3. The variable `r` can have any real value, and `w` is restricted to be binary (that is 0/1 or True/False). If the domain is not specified, the default is the `Reals` virtual set. Other virtual sets supported by Pyomo are defined in Table 4.2. Note that these virtual sets can also be used in other contexts (e.g., when constructing `Param` objects).

`Any`	The set of all possible values, except `None`
`AnyWithNone`	The set of all possible values
`EmptySet`	The set with no data values
`Reals`	The set of floating point values
`PositiveReals`	The set of strictly positive floating point values
`NonPositiveReals`	The set of non-positive floating point values
`NegativeReals`	The set of strictly negative floating point values
`NonNegativeReals`	The set of non-negative floating point values
`PercentFraction`	The set of floating point values in the interval [0,1]
`UnitInterval`	The same as 'PercentFraction'
`Integers`	The set of integer values
`PositiveIntegers`	The set of positive integer values
`NonPositiveIntegers`	The set of non-positive integer values
`NegativeIntegers`	The set of negative integer values
`NonNegativeIntegers`	The set of non-negative integer values
`Boolean`	The set of boolean values, which can be represented as False/True, 0/1, 'False'/'True' and 'F'/'T'
`Binary`	The same as 'Boolean'

Table 4.2: Predefined virtual sets in Pyomo.

The `domain` or `within` argument can also accept a function, which is used to define the domain for individual elements of an indexed variable. For example:

```
model.A = pyo.Set(initialize=[1,2,3])
def s_domain(model, i):
    return pyo.RangeSet(i, i+1, 1) # (start, end, step)
model.s = pyo.Var(model.A, domain=s_domain)
```

In this example, `s` is an indexed variable whose individual entities are defined over consecutive integer intervals.

> **NOTE:** While Pyomo supports a general representation for restricting the domain of the variables, not all solvers support this general behavior. You may need to restrict your definitions to those supported by the selected solver.

Variable bounds can be explicitly specified with the bounds keyword option:

```
model.A = pyo.Set(initialize=[1,2,3])
model.a = pyo.Var(bounds=(0.0,None))

lower = {1:2.5, 2:4.5, 3:6.5}
upper = {1:3.5, 2:4.5, 3:7.5}
def f(model, i):
    return (lower[i], upper[i])
model.b = pyo.Var(model.A, bounds=f)
```

The bounds option can specify a 2-tuple with lower and upper values. Alternatively, it can specify a function that returns a 2-tuple for each variable index. Note that None can be used in place of the lower or upper bound to indicate no bound should be enforced. In the code snippet above, model.a has a lower bound of 0, and does not have an upper bound, while model.b has different bounds for each of its indices. For example, model.b[3] has a lower bound of 6.5 and an upper bound of 7.5.

The initial value of variables can be set with the initialize keyword argument as in the following example:

```
model.A = pyo.Set(initialize=[1,2,3])
model.za = pyo.Var(initialize=9.5, within=pyo.NonNegativeReals)
model.zb = pyo.Var(model.A, initialize={1:1.5, 2:4.5, 3:5.5})
model.zc = pyo.Var(model.A, initialize=2.1)

print(pyo.value(model.za)) # 9.5
print(pyo.value(model.zb[3])) # 5.5
print(pyo.value(model.zc[3])) # 2.1
```

For non-indexed variables, a single scalar value is provided to the initialize keyword argument. If the component is indexed, a single value can still be provided, in which case all entries in an indexed variable will be initialized to the same value. As well, a dictionary can be passed in where the keys correspond to the valid indices of the variable. Additionally, this argument can be passed a rule (a Python function) that accepts the model and variable indices as arguments and returns the desired initial value for that variable element:

```
model.A = pyo.Set(initialize=[1,2,3])
def g(model, i):
    return 3*i
model.m = pyo.Var(model.A, initialize=g)

print(pyo.value(model.m[1])) # 3
print(pyo.value(model.m[3])) # 9
```

4.3.2 Working with `Var` Objects

When generating formatted output, or scripting advanced workflows, there are several attributes and methods of `Var` commonly used. Consider the following declarations:

```
model.A = pyo.Set(initialize=[1,2,3])
model.za = pyo.Var(initialize=9.5, within=pyo.NonNegativeReals)
model.zb = pyo.Var(model.A, initialize={1:1.5, 2:4.5, 3:5.5})
model.zc = pyo.Var(model.A, initialize=2.1)
```

The current value of the variable can be obtained with the `value()` function, and the attributes `lb` and `ub` hold values for the lower and upper bounds on the variable, respectively. These values may be inferred from the domain of the variable.

```
print(pyo.value(model.zb[2])) # 4.5
print(model.za.lb) # 0
print(model.za.ub) # None
```

The `setlb` and `setub` methods are used to set the lower and upper bounds on a variable.

Variable values can be set using the Python assignment operator,

```
model.za = 8.5
model.zb[2] = 7.5
```

One can also call the `set_values` method to set all the variable values from a dictionary.

`Var` components can be fixed to specific values. If the `fixed` attribute is `True`, then the variable has a fixed value that will not be altered by an optimizer. The `fix` method is used to fix elements of a `Var`, and the `unfix` method is used to unfix elements of a `Var`.

```
model.zb.fix(3.0)
print(model.zb[1].fixed) # True
print(model.zb[2].fixed) # True
model.zc[2].fix(3.0)
print(model.zc[1].fixed) # False
print(model.zc[2].fixed) # True
```

4.4 Objectives

An objective is a function that is either minimized or maximized by a solver. The solver searches for values of the variables that result in the best possible value of the objective function. The following sections describe the syntax for declaring and working with objectives.

4.4.1 *Objective Declarations*

Most solvers can be applied to optimization models with a single objective. The following code creates an `Objective` object:

```
model.a = pyo.Objective()
```

Named and un-named arguments are supported, and Table 4.3 provides a list of the common arguments that can be passed when declaring the `Objective` component.

keyword	description	acceptable values
<un-named>	reserved for specifying index sets	any number of Pyomo `Set` objects or Python lists
expr	provides the expression that defines the objective function	any valid Pyomo expression
rule	provides the rule function that will be called to access the expression that defines the objective function	a function that returns a Pyomo expression or `Objective.Skip`
sense	determines if the objective is to be minimized or maximized (default is to minimize)	`minimize` or `maximize`

Table 4.3: Common declaration arguments for the `Objective` component

The `expr` keyword can be used to specify the actual expression for the objective. One can also use the `rule` keyword to specify a rule (a Python function) that returns an expression. A rule provides control over how the objective is formed. Both options are illustrated here:

```
model.x = pyo.Var([1,2], initialize=1.0)

model.b = pyo.Objective(expr=model.x[1] + 2*model.x[2])

def m_rule(model):
    expr = model.x[1]
    expr += 2*model.x[2]
    return expr
model.c = pyo.Objective(rule=m_rule)
```

Some solvers can perform multi-objective optimization with two or more objectives. Multiple objectives can be declared individually or they can be indexed and defined using a rule as shown here:

```
A = ['Q', 'R', 'S']
model.x = pyo.Var(A, initialize=1.0)
def d_rule(model, i):
    return model.x[i]**2
model.d = pyo.Objective(A, rule=d_rule)
```

When the Objective object is declared as an indexed component, Pyomo iterates over all elements of the index set during object construction, passing each set element to the function given as the argument to the rule keyword. If multiple sets are specified in an Objective declaration, then Pyomo iterates over the cross product of all sets, providing an element for each set to the rule function.

In some contexts, it may be convenient to not define objectives for some index values. If the construction rule returns Objective.Skip, then the objective is ignored.

```
def e_rule(model, i):
    if i == 'R':
        return pyo.Objective.Skip
    return model.x[i]**2
model.e = pyo.Objective(A, rule=e_rule)
```

By default, the declaration of an Objective object indicates that the objective is to be minimized. The sense keyword can be used to indicate an objective that is maximized using sense=pyo.maximize

4.4.2 Working with Objective Objects

The objective function contains a few attributes that may be useful for scripting or debugging. The expr attribute stores the expression for the objective. The sense attribute indicates if the objective is to be minimized or maximized. The value function can be used to compute the value of the objective. These are illustrated in the following example:

```
A = ['Q', 'R']
model.x = pyo.Var(A, initialize={'Q':1.5, 'R':2.5})
model.o = pyo.Objective(expr=model.x['Q'] + 2*model.x['R'])
print(model.o.expr) # x[Q] + 2*x[R]
print(model.o.sense) # minimize
print(pyo.value(model.o)) # 6.5
```

4.5 Constraints

A constraint defines one or more expressions that place limits on the feasible values of variables. The declaration of constraint expressions is similar to the declaration of objective function expressions. Constraints differ from objectives in that the expressions include relationships (equalities or inequalities). While objectives can be indexed, this feature is infrequently used. In contrast, constraints are commonly indexed, allowing for a collection of related constraint expressions to be constructed and stored in a single constraint object.

4.5.1 `Constraint` Declarations

The following code creates a single, non-indexed `Constraint` object:

```
model.x = pyo.Var([1,2], initialize=1.0)
model.diff = pyo.Constraint(expr=model.x[2]-model.x[1] <= 7.5)
```

Several named arguments are supported, and Table 4.4 lists the common arguments that can be passed when declaring a `Constraint` component.

The expression specified by the `expr` keyword can alternatively be generated with a rule function. For example, the `diff` constraint can also be declared as follows:

```
model.x = pyo.Var([1,2], initialize=1.0)
def diff_rule(model):
    return model.x[2] - model.x[1] <= 7.5
model.diff = pyo.Constraint(rule=diff_rule)
```

keyword	description	acceptable values
<un-named>	reserved for specifying index sets	any number of Pyomo `Set` objects or Python lists
`expr`	provides the expression that defines the constraint	any valid Pyomo expression with a relational operator, a 2-tuple, or a 3-tuple
`rule`	provides the rule function that will be called to access the expression that defines the constraint	a function that returns a Pyomo expression with a relational operator, a 2-tuple, a 3-tuple, or `Constraint.Skip`

Table 4.4: Common declaration arguments for the `Constraint` component

Constraints can be indexed, and those indices can be used to refer to specific elements of indexed parameters and variables when constructing expressions. The following code fragment shows an example of this:

```
N = [1,2,3]

a = {1:1, 2:3.1, 3:4.5}
b = {1:1, 2:2.9, 3:3.1}

model.y = pyo.Var(N, within=pyo.NonNegativeReals, initialize=0.0)

def CoverConstr_rule(model, i):
    return a[i] * model.y[i] >= b[i]
model.CoverConstr = pyo.Constraint(N, rule=CoverConstr_rule)
```

Indexed constraints are specified in the same manner as indexed objectives. Pyomo iterates over the cross product of the indexing sets, providing an index from each set to the rule function. The CoverConstr constraint in this example implements the following mathematical model:

$$a_i y_i \geq b_i \ \forall i \in \{1,2,3\} \tag{4.1}$$

Given the data specified in a and b, the model instance passed to the solver will include the following explicit constraints:

$$y[1] \geq 1$$
$$3.1 \cdot y[2] \geq 2.9$$
$$4.5 \cdot y[3] \geq 3.1$$

Three types of constraint expressions are allowed in Pyomo:

- *inequality constraints* have the form

$$expr_1 \leq expr_2 \quad \text{or} \quad expr_1 \geq expr_2$$

where $expr_1$ and $expr_2$ may be non-constant expressions. (Note that $<$ and $>$ are not supported.)
- *equality constraints* have the form

$$expr_1 = expr_2$$

where $expr_1$ and $expr_2$ may be non-constant expressions.
- *range constraints* have the form

$$lower \leq expr_1 \leq upper$$

where *lower* and *upper* are constant expressions and $expr_1$ is a non-constant expression.

In some optimization models, a constraint might not be defined for all indices. For example, particular indices might not be physically realizable. The rule function can return `Constraint.Skip` (or `Constraint.NoConstraint`) to indicate that no constraint is associated with a particular index. For example, consider the declaration of a notional task scheduling constraint:

```
TimePeriods = [1,2,3,4,5]
LastTimePeriod = 5

model.StartTime = pyo.Var(TimePeriods, initialize=1.0)

def Pred_rule(model, t):
    if t == LastTimePeriod:
        return pyo.Constraint.Skip
    else:
        return model.StartTime[t] <= model.StartTime[t+1]

model.Pred = pyo.Constraint(TimePeriods, rule=Pred_rule)
```

The value `Constraint.Skip` indicates that no constraint is being generated, and the corresponding index value is skipped. An alternative to this approach is to construct a sparse index set that specifies only the valid indices in the constraint. However, this may not always be practical in complex models (for a discussion of sparse index sets, see Section 9.4).

The value `Constraint.Feasible` indicates that the constraint generated for the specified index is always feasible. Consequently, that constraint does not need to be generated, and it is skipped. Similarly, the value `Constraint.Infeasible` indicates that the constraint generated by the specified index is infeasible. This might be used, for example, if a particular combination of parameter values produced an invalid constraint. For this value, Pyomo raises an exception to inform the user, because this typically indicates an error in the model or data.

4.5.2 *Working with* `Constraint` *Objects*

After a constraint is declared, the constraint expression is processed to identify the elements of the logical tuple: (*lower*, *body*, *upper*), where the non-constant expressions are pushed to the body. Hence, the `lower` and `upper` attributes are constant expressions or `None`, and the `body` attribute contains a Pyomo expression. If a `Constraint` contains an equality expression, then the `equality` attribute is `True`, and the `lower` and `upper` attributes have the same value.

The value of the constraint body can be evaluated using the `value` function. Similarly, the `lslack` and `uslack` methods can be used to compute slack values (the difference between the current expression value and the lower or upper bound), as shown in the following example:

```
model = pyo.ConcreteModel()
model.x = pyo.Var(initialize=1.0)
model.y = pyo.Var(initialize=1.0)

model.c1 = pyo.Constraint(expr=model.y - model.x <= 7.5)
model.c2 = pyo.Constraint(expr=-2.5 <= model.y - model.x)
model.c3 = pyo.Constraint(
    expr=pyo.inequality(-3.0, model.y - model.x, 7.0))

print(pyo.value(model.c1.body)) # 0.0

print(model.c1.lslack()) # inf
print(model.c1.uslack()) # 7.5
print(model.c2.lslack()) # 2.5
print(model.c2.uslack()) # inf
print(model.c3.lslack()) # 3.0
print(model.c3.uslack()) # 7.0
```

4.6 Set Data

A set is a collection of data, possibly including numeric data (e.g., real or integer values) as well as symbolic data (e.g., strings) typically used to specify the valid indices for an indexed component. Several classes can be used to define sets in Pyomo models:

Set	A generic component for declaring sets
RangeSet	A component that defines a range of numbers
SetOf	A component that creates a set from external data without copying the data

4.6.1 Set Declarations

The following code creates a `Set` object:

```
model.A = pyo.Set()
```

Named and un-named arguments are supported, and Table 4.5 provides a list of the common arguments that can be passed when declaring the `Set` component.

keyword	description	acceptable values
<un-named>	reserved for specifying index sets	any number of Pyomo `Set` objects or Python lists
`initialize`	provides initial values to store in the set	Python list, Python dictionary, or rule function
`within` `domain`	specifies the valid values that can be contained in the set	a Pyomo `Set` object or Python list
`ordered`	specifies whether or not order of the set should be preserved	True, False, `Set.InsertionOrder`, or `Set.SortedOrder`
`bounds`	specifies the lower and upper bounds for valid values in the set	a 2-tuple, a Python dictionary, or a rule function
`filter`	specifies a rule for determining membership in the set	a rule function

Table 4.5: Common declaration arguments for the `Set` component

An indexed set can also be specified by providing other sets or Python lists as un-named arguments in the declaration:

```
model.A = pyo.Set()
model.B = pyo.Set()
model.C = pyo.Set(model.A)
model.D = pyo.Set(model.A,model.B)
```

Similarly, standard Python types can be used to define a set index:

```
model.E = pyo.Set([1,2,3])
f = set([1,2,3])
model.F = pyo.Set(f)
```

Set declarations can also use standard set operations to declare a set in a constructive fashion:

```
model.A = pyo.Set()
model.B = pyo.Set()
model.G = model.A | model.B # set union
model.H = model.B & model.A # set intersection
model.I = model.A - model.B # set difference
model.J = model.A ^ model.B # set exclusive-or
```

Also, set cross-products can be specified with the multiplication operator:

```
model.A = pyo.Set()
model.B = pyo.Set()
model.K = model.A * model.B
```

The `initialize` keyword can be used to specify the elements in a set:

```
model.B = pyo.Set(initialize=[2,3,4])
model.C = pyo.Set(initialize=[(1,4),(9,16)])
```

A Python dictionary can also be passed to the `initialize` keyword to specify the elements for each index of an indexed set:

```
F_init = {}
F_init[2] = [1,3,5]
F_init[3] = [2,4,6]
F_init[4] = [3,5,7]
model.F = pyo.Set([2,3,4],initialize=F_init)
```

Alternatively, a rule (a Python function) can be passed to the `initialize` keyword to provide the elements for an indexed set. The function accepts the model and indices and returns the desired set for that index:

```
def J_init(model, i, j):
    return range(0,i*j)
model.J = pyo.Set(model.B,model.B, initialize=J_init)
```

The previous examples illustrate how data can be specified or dynamically generated to initialize a set. There are some contexts where it is simpler to specify the set elements that should be omitted. The `filter` keyword can be used to specify a function that returns `True` when an element belongs in a set, and `False` otherwise. For example:

```
model.P = pyo.Set(initialize=[1,2,3,5,7])
def filter_rule(model, x):
    return x not in model.P
model.Q = pyo.Set(initialize=range(1,10), filter=filter_rule)
```

Here, set P contains prime values, and set Q is the set of all numbers except for the members of P.

After an indexed set is constructed in a concrete model, sets can be added for specific indices using the Python equal operator:

```
model.R = pyo.Set([1,2,3])
model.R[1] = [1]
model.R[2] = [1,2]
```

Validation of set data is supported in two different ways. First, a superset can be specified with the `within` or `domain` keyword:

```
model.B = pyo.Set(within=model.A)
```

When an element is added to the set B, it is checked to confirm that it also belongs to A. This ensures B is a subset of A.

Validation of set data can also be performed by passing a rule to the `validate` keyword argument. The rule function should return `True` if the element that is passed in belongs in this set, and `False` otherwise (Pyomo will throw an exception). For example, the following C_validate function mimics the `within` keyword argument:

```
def C_validate(model, value):
    return value in model.A
model.C = pyo.Set(validate=C_validate)
```

Finally, note that if both the `within` and `validate` keyword arguments are specified, then the logic specified by both are applied to validate set elements.

By default, sets are ordered by insertion order. In some cases, we may want the set elements to be in sorted order. This can be done using the `Set.SortedOrder` option with the `ordered` keyword:

```
model.A = pyo.Set(ordered=pyo.Set.SortedOrder)
```

Sets may contain data elements that are either singletons or k-tuples. The `dimen` keyword is used to specify the expected dimension of the data. The default value is one, indicating the set will contain singleton data. In some cases, the appropriate value of the dimension can be determined from other keyword values, but in general the user is required to specify this keyword for tuple set data.

Ordered sets may have first and last values. The `bounds` option can be used to specify a 2-tuple defining upper and lower bounds for a set. This option may be inferred from the `within` argument, when the set is ordered.

The `RangeSet` component defines an ordered virtual set that represents a sequence of integer or floating point values. This sequence is defined by a start value, a final value, and a step size. If a `RangeSet` is defined with a single argument, then the argument defines the final value. The start value defaults to 1 and the step size defaults to 1. For example, the following defines a sequence of integers from 1 to 10:

```
model.A = pyo.RangeSet(10)
```

If a `RangeSet` is defined with two arguments, then the first is the start value and the second is the final value. For example, the following defines a sequence of integers from 5 to 10:

```
model.C = pyo.RangeSet(5,10)
```

Finally, if a `RangeSet` is defined with three arguments, then they are the start value, final value and step size respectively. For example, the following defines a sequence of floating point values from 2.5 to 10.0 with step 1.5:

```
model.D = pyo.RangeSet(2.5,11,1.5)
```

4.6.2 Working with Set Objects

The `len()` function returns the number of elements in the set:

```
model.A = pyo.Set(initialize=[1,2,3])

print(len(model.A)) # 3
```

The elements in the set can be accessed with the data() method, which returns the underlying set data as a Python tuple (or a Python dictionary for indexed sets) as shown below:

```python
model.A = pyo.Set(initialize=[1, 2, 3])
model.B = pyo.Set(initialize=[3, 2, 1], ordered=True)
model.C = pyo.Set(model.A, initialize={1:[1], 2:[1, 2]})

print(type(model.A.data()) is tuple) # True
print(type(model.B.data()) is tuple) # True
print(type(model.C.data()) is dict) # True
print(sorted(model.A.data())) # [1, 2, 3]
for index in sorted(model.C.data().keys()):
  print(sorted(model.C.data()[index]))
# [1]
# [1, 2]
```

Set comparison and membership tests are supported with a variety of Python operators:

```python
model.A = pyo.Set(initialize=[1,2,3])

# Test if an element is in the set
print(1 in model.A) # True

# Test if sets are equal
print([1, 2] == model.A) # False

# Test if sets are not equal
print([1, 2] != model.A) # True

# Test if a set is a subset of or equal to the set
print([1, 2] <= model.A) # True

# Test if a set is a subset of the set
print([1, 2] < model.A) # True

# Test if a set is a superset of the set
print([1, 2, 3] > model.A) # False

# Test if a set is a superset of or equal to the set
print([1, 2, 3] >= model.A) # True
```

Sets can also be iterated over to access individual elements in the set:

```
model.A = pyo.Set(initialize=[1, 2, 3])
model.C = pyo.Set(model.A, initialize={1:[1], 2:[1, 2]})

print(sorted(e for e in model.A)) # [1, 2, 3]
for index in model.C:
    print(sorted(e for e in model.C[index]))
# [1]
# [1, 2]
```

Ordered sets include a variety of methods that reflect the ordering in the set:

```
model.A = pyo.Set(initialize=[3, 2, 1], ordered=True)

print(model.A.first()) # 3
print(model.A.last()) # 1
print(model.A.next(2)) # 1
print(model.A.prev(2)) # 3
print(model.A.nextw(1)) # 3
print(model.A.prevw(3)) # 1
```

The `first()` and `last()` methods return the first and last elements in an ordered set respectively. The `next()` method takes an element in the set and returns the next element in the set. Similarly, the `prev()` method returns the previous element. The `nextw()` and `prevw()` methods operate similarly, except that they wrap around the ends of the set. In this example, the value of `nextw(1)` is 3 because 1 is the last element of the set, and 3 is the next element if the set indices wrap around. The `ord()` method can be used to find the position index of an element in an ordered set, and the `[]` operator can be used to access an element given a position index:

```
model.A = pyo.Set(initialize=[3, 2, 1], ordered=True)

print(model.A.ord(3)) # 1
print(model.A.ord(1)) # 3
print(model.A[1]) # 3
print(model.A[3]) # 1
```

NOTE: The position indices start at one, not zero. The order of the set is determined by the sequence of the data provided when it is instantiated and the option specified for the `ordered` keyword argument.

4.7 Parameter Data

A parameter is a numerical or symbolic value used to formulate constraints and objectives in a model. Pyomo parameters can be created using the `Param` class, which can denote a single value, an array of values, or a multi-dimensional array of values.

An unindexed `Param` component looks a lot like a scalar value, and an indexed `Param` component looks a lot like a Python dictionary of values. The `Param` component supports advanced features like mutability and sparse representations with default values.

4.7.1 *Param Declarations*

The following code creates a `Param` object:

```
model.Z = pyo.Param(initialize=32)
```

Named and un-named arguments are supported, and Table 4.6 provides a list of the common arguments that can be passed when declaring the `Param` component.

keyword	description	acceptable values
<un-named>	reserved for specifying index sets	any number of Pyomo `Set` objects or Python lists
initialize	provides an initial value for the parameter	a scalar value, Python dictionary, or rule function
default	provides default a value to use for the parameter if no value has been set	a scalar value, Python dictionary, or rule function
validate	specifies a function that is called to determine if a particular value is valid for the parameter	a function that returns True or False given a particular value
mutable	specifies whether or not the parameter value may change between calls to a solver	True/False

Table 4.6: Common declaration arguments for the `Param` component

An indexed parameter can be specified by providing sets as un-named arguments to the `Param` declaration:

```
model.A = pyo.Set(initialize=[1,2,3])
model.B = pyo.Set(initialize=['A','B'])
model.U = pyo.Param(model.A, initialize={1:10, 2:20, 3:30})
model.T = pyo.Param(model.A, model.B,
                initialize={(1,'A'):10, (2,'B'):20, (3,'A'):30})
```

The `initialize` keyword can be used to specify the value of a parameter as shown in the previous two code snippets. A rule function can also be passed to the `initialize` keyword to set the value of a parameter:

```
def X_init(model, i, j):
    return i*j
model.X = pyo.Param(model.A, model.A, initialize=X_init)
```

If ordered sets are used to define the index for an indexed parameter, then the initialization function can reference previously defined parameter values:

```
def XX_init(model, i, j):
    if i==1 or j==1:
        return i*j
    return i*j + model.XX[i-1,j-1]
model.XX = pyo.Param(model.A, model.A, initialize=XX_init)
```

The `default` option can be used to specify parameter values for all valid indices that have not been explicitly initialized. For example, we can define an indexed parameter that represents a 3×3 diagonal matrix as follows:

```
u={}
u[1,1] = 10
u[2,2] = 20
u[3,3] = 30
model.U = pyo.Param(model.A, model.A, initialize=u, default=0)
```

Similar to the `Set` component, there are two ways to validate parameter values. First, the `within` keyword option can be used to specify the valid domain of parameter values:

```
model.Z = pyo.Param(within=pyo.Reals)
```

Validation of parameter data can also be performed with the `validate` option, which specifies a function that returns `True` if a parameter value is valid and `False` if it is not (Pyomo will throw an exception). The following example uses the `validate` option to mimic the behavior of the `within` option:

```
def Y_validate(model, value):
    return value in pyo.Reals
model.Y = pyo.Param(validate=Y_validate)
```

Validation of indexed parameters is performed similarly. The `validate` option specifies a function whose arguments are the model, parameter value, and the parameter indices:

```
model.A = pyo.Set(initialize=[1,2,3])
def X_validate(model, value, i):
    return value > i
model.X = pyo.Param(model.A, validate=X_validate)
```

If both the `within` and `validate` options are specified, then the logic for both of these options will be applied to validate parameter values.

The `Param` component can be used to represent constant values in Pyomo models; however, mutability is also supported. In the following example, Pyomo generates the expression for the objective in this model with the form:

$$x_1 + 4x_2 + 9x_3.$$

Specifically, Pyomo has treated parameter values as fixed constants, and its expressions simply contain the numeric constants.

```
model = pyo.ConcreteModel()
p = {1:1, 2:4, 3:9}

model.A = pyo.Set(initialize=[1,2,3])
model.p = pyo.Param(model.A, initialize=p)
model.x = pyo.Var(model.A, within=pyo.NonNegativeReals)

model.o = pyo.Objective(expr=sum(model.p[i]*model.x[i] for i in \
    model.A))
```

Note that this "conversion" happens as soon as the expression is first created. The fact that these values come from a `Param` component is lost, and only the numerical values remain. This is done for efficiency. Consequently, these values *cannot be changed* once the expression is created.

However, this behavior is different if the `mutable` option is specified while constructing the model. If this option is `True`, then the parameter values are not treated as constants. Consider the previous example again where the p parameter is now mutable:

```
model = pyo.ConcreteModel()
p = {1:1, 2:4, 3:9}

model.A = pyo.Set(initialize=[1,2,3])
model.p = pyo.Param(model.A, initialize=p, mutable=True)
model.x = pyo.Var(model.A, within=pyo.NonNegativeReals)

model.o = pyo.Objective(expr=pyo.summation(model.p, model.x))

model.p[2] = 4.2
model.p[3] = 3.14
```

When Pyomo generates the expression for the objective in this model, it keeps knowledge of the `Param` component and now has the form:

$$p_1 x_1 + p_2 x_2 + p_3 x_3,$$

where the values p_i are `Param` objects with references to the parameter values. Here, Pyomo treats the parameter values as mutable values that may be changed later by the user. In this example, the parameter values are changed *after* the objective expression is defined, and the resulting objective is

$$x_1 + 4.2x_2 + 3.14x_3.$$

The parameters are only replaced with their numerical values when calling the solver. Therefore, their values can be changed between consecutive calls to a solver.

Mutable parameters require some additional overhead for memory and they require additional processing when translating Pyomo expressions into a form that a solver understands. Consequently, parameters are immutable by default.

4.7.2 Working with Param Objects

Pyomo assumes parameter values are specified with a sparse representation. For example, the Param object T declares a parameter indexed over sets A and B:

```
model.T = pyo.Param(model.A, model.B)
```

However, not all of these values are required to be defined in a model. For example:

```
model.B = pyo.Set(initialize=[1,2,3])
w={}
w[1] = 10
w[3] = 30
model.W = pyo.Param(model.B, initialize=w)
```

Parameter W is defined for indices 1 and 3, but the index set B includes 1, 2, and 3. If W[2] is accessed, an error occurs and a Python exception is thrown.

As mentioned earlier, a default value can also be provided with the default keyword argument. If a default value is provided, and a model tries to access a value that has not been initialized, the default value is used (instead of throwing an exception). Note that the parameter data is stored with a sparse representation, even if the default value is specified. This is supported for memory efficiency. It provides a convenient way for the modeler to reference sparse values without adopting a specialized data structure.

Because of this sparse representation, several methods that consider the valid keys of an indexed parameter require specialized behavior. Let the *valid index set* refer to the complete list of all valid indices whether initialized or not, and let the *effective index set* denote only the set of initialized key values in an indexed component. If no default value is declared, then the the len function returns the size of the effective index set, and the in operator tests if a specified value is in the effective index set. Iteration is supported over values in the effective index set, and the Python [] operator can be used to access individual elements (which is the parameter value in this example).

If a default value is declared, then all indices are equally valid in the model, whether explicitly indexed or not. Therefore, the len() function returns the size of the full index set, iteration and the in operator consider the full index set. Thus, when a default value is specified, the parameter appears to be densely populated with values, even if the underlying data structure is kept sparse for efficiency. This is illustrated in the following example:

```
model = pyo.ConcreteModel()
model.p = pyo.Param([1,2,3], initialize={1:1.42, 3:3.14})
model.q = pyo.Param([1,2,3], initialize={1:1.42, 3:3.14}, \
    default=0)
```

```
# Demonstrating the len() function
print(len(model.p))  # 2
print(len(model.q))  # 3

# Demonstrating the 'in' operator (checks against component keys)
print(2 in model.p)  # False
print(2 in model.q)  # True

# Demonstrating iteration over component keys
print([key for key in model.p])  # [1, 3]
print([key for key in model.q])  # [1, 2, 3]
```

The methods `sparse_keys()`, `sparse_values()`, `sparse_items()`, `sparse_iterkeys()`, `sparse_itervalues()`, and `sparse_iteritems()` define sparse versions of the corresponding methods defined in the `IndexedComponent` class. These methods return values only for the defined parameter values, whether or not a default value is specified.

4.8 Named Expressions

Pyomo expressions are mathematical statements containing numbers, parameters, and variables combined using operators such as $+, -, *, /$ for example. These expressions form the basis of the algebraic representation of a model, and are stored inside constraint and objective components on the model.

The `Expression` component provides a mechanism for storing a Pyomo expression on a model making the expression reusable in multiple contexts, such as a common sub-expression in one or more constraints, without the overhead of regenerating the expression each time. In addition, the Pyomo expression stored by the `Expression` component can be changed at a later time, thereby updating any constraint or objective expressions referencing it. This provides a powerful approach for modifying a model between calls to a solver.

The following sections describe the syntax for declaring and working with named expressions.

4.8.1 *Expression Declarations*

The following code creates a single, non-indexed `Expression` object:

```
model.e = pyo.Expression()
```

Named and un-named arguments are supported, and Table 4.7 provides a list of the common arguments that can be passed when declaring the `Expression` component.

The `expr` or `rule` keywords can be used to initialize a named expression when it is declared, as shown in the following example:

```
model.x = pyo.Var()
model.e1 = pyo.Expression(expr=model.x + 1)
def e2_rule(model):
    return model.x + 2
model.e2 = pyo.Expression(rule=e2_rule)
```

keyword	description	acceptable values
<un-named>	reserved for specifying index sets	any number of Pyomo Set objects or Python lists
expr	provides the expression to store	any valid Pyomo expression
rule	provides the rule function that will be called to provide the expression to store	a function that returns a Pyomo expression or Expression.Skip

Table 4.7: Common declaration arguments for the Expression component

As with the other core modeling components, the Expression component can be indexed by declaring it with one or more unnamed arguments representing indexing sets. The following example declares an indexed Expression component over all members of the index set except for the first. Indices that should be left out of the indexed Expression component are signified by returning the Expression.Skip attribute from the rule function.

```
N = [1,2,3]
model.x = pyo.Var(N)
def e_rule(model, i):
    if i == 1:
        return pyo.Expression.Skip
    else:
        return model.x[i]**2
model.e = pyo.Expression(N, rule=e_rule)
```

4.8.2 Working with Expression Objects

A simple use for the Expression component declares a single expression and uses it inside an objective and a constraint declaration:

```
model.x = pyo.Var()
model.e = pyo.Expression(expr=(model.x - 1.0)**2)
model.o = pyo.Objective(expr=0.1*model.e + model.x)
model.c = pyo.Constraint(expr=model.e <= 1.0)
```

The value of the named expression can be computed using the `value` function. Additionally, the expression stored in the named Expression component can be updated. As the following example shows, updating the named expression has the effect of updating the objective and constraint expressions where it is used:

```
model.x.set_value(2.0)
print(pyo.value(model.e))  # 1.0
print(pyo.value(model.o))  # 2.1
print(pyo.value(model.c.body))  # 1.0

model.e.set_value((model.x - 2.0)**2)
print(pyo.value(model.e))  # 0.0
print(pyo.value(model.o))  # 2.0
print(pyo.value(model.c.body))  # 0.0
```

The `Expression` component does not require an expression when it is declared on a model, but it must be assigned one before the model is solved if the named expression is used in any active objectives or constraints. Furthermore, named expressions that are used in objectives or constraints should not store relational Pyomo expressions, that is, expressions using one or more of the operators <=, <, >=, >, and ==.

4.9 Suffix Components

Suffixes provide a mechanism for annotating a model with auxiliary data not strictly related to the model declaration and structure. Suffixes are commonly used by solver plugins to store extra information about the solution of a model. More generally, suffixes can be used to

- import information from a solver about the solution to a mathematical program (e.g., constraint duals, variable reduced costs, basis information),
- export information to a solver or algorithm to configure the solution process (e.g., warm-starting information, variable branching priorities), and
- tag model components with local data for later use in advanced scripting algorithms.

This functionality is made available to the modeler through the `Suffix` component class, providing an interface for annotating Pyomo modeling components with additional data.

4.9.1 *Suffix Declarations*

The following code creates a suffix labeled `foo`:

```
model.foo = pyo.Suffix()
```

Named arguments are supported, and Table 4.8 provides a list of the common arguments that can be passed when declaring the `Suffix` component

keyword	description	acceptable values
direction	specifies if a suffix is an input to or an output from a solver	Suffix.LOCAL, Suffix.IMPORT, Suffix.EXPORT, Suffix.IMPORT_EXPORT (more details given below)
datatype	specifies the particular type of data being stored in the suffix	Suffix.FLOAT, Suffix.INT, None (more details given below
initialize	provides initial values for the suffix	a rule function

Table 4.8: Common declaration arguments for the `Suffix` component

The `Suffix` component is a not an indexed component, and hence it cannot be declared with unnamed positional arguments. The `direction` keyword argument is used to specify the information flow for a suffix when interfacing with a solver. This argument can be one of four possible values:

- `Suffix.LOCAL`: Suffix data is local to the model. It is not imported or exported by solver plugins and is the default.
- `Suffix.IMPORT`: Suffix data will be imported from solvers to the model by solver plugins.
- `Suffix.EXPORT`: Suffix data will be exported from the model to the solver by the plugins.
- `Suffix.IMPORT_EXPORT`: Suffix data is both imported and exported by solver plugins.

Not all solver plugins are guaranteed to manage suffix information flow, but the user controls this information flow by configuring suffix components.

The `datatype` keyword argument specifies the type of data held in the suffix. This argument can be one of three possible values:

- `Suffix.FLOAT`: floating point data (default).
- `Suffix.INT`: integer data.
- `None`: any type of data.

This argument might be optional for some solver interfaces; however, exporting suffix data with solvers using the `nl` file interface requires all active export suffixes have a strict datatype (i.e., the `datatype` keyword cannot be `None`). The following example illustrates various suffix declarations:

```
# Export integer data
model.priority = pyo.Suffix(direction=pyo.Suffix.EXPORT,
                datatype=pyo.Suffix.INT)
```

```
# Export and import floating point data
model.dual = pyo.Suffix(direction=pyo.Suffix.IMPORT_EXPORT)
```

Suffixes are not guaranteed to be compatible with all solver plugins in Pyomo. Whether a given suffix is acceptable or not depends on both the solver and solver interface being used. In some cases, a solver plugin will raise an exception if it encounters a suffix type that it does not handle, but this is not true in every situation. For example, the nl file interface is generic to all AMPL-compatible solvers, so there is no way for Pyomo to validate that a suffix of a given name, direction, and datatype is appropriate for a solver. One should be careful in verifying that suffix declarations are being handled as expected when switching to a different solver or solver interface.

The initialize keyword argument can be used to define suffix values. This argument specifies a function that is executed when the model is constructed. This function returns a list or iterable of (component, value) tuples.

```
model = pyo.AbstractModel()
model.x = pyo.Var()
model.c = pyo.Constraint(expr=model.x >= 1)

def foo_rule(m):
  return ((m.x, 2.0), (m.c, 3.0))
model.foo = pyo.Suffix(initialize=foo_rule)
```

4.9.2 Working with Suffixes

Consider the following example.

```
model = pyo.ConcreteModel()
model.x = pyo.Var()
model.y = pyo.Var([1,2,3])
model.foo = pyo.Suffix()
```

This examples includes two variable components, indexed and non-indexed, along with a suffix component. Conceptually, the declaration of the suffix foo allows the association of foo with each component in the model. For example:

```
# Assign the value 1.0 to suffix 'foo' for model.x
model.x.set_suffix_value('foo', 1.0)

# Assign the value 2.0 to suffix model.foo for model.x
model.x.set_suffix_value(model.foo, 2.0)

# Get the value of suffix 'foo' for model.x
print(model.x.get_suffix_value('foo')) # 2.0
```

Suffix values can be assigned with set_suffix_value and they can be accessed with get_suffix_value. This example illustrates two ways of specifying the same suffix: with a name and with a suffix component object.

Suffix values for indexed components can also be assigned with
`set_suffix_value`:

```
# Assign the value 3.0 to suffix model.foo for model.y
model.y.set_suffix_value(model.foo, 3.0)

# Assign the value 4.0 to suffix model.foo for model.y[2]
model.y[2].set_suffix_value(model.foo, 4.0)

# Get the value of suffix 'foo' for model.y
print(model.y.get_suffix_value(model.foo))    # None
print(model.y[1].get_suffix_value(model.foo)) # 3.0
print(model.y[2].get_suffix_value(model.foo)) # 4.0
print(model.y[3].get_suffix_value(model.foo)) # 3.0
```

This example illustrates how `set_suffix_value` is used to set the value for an
indexed component and a single component data object. When
`set_suffix_value` is called for an indexed component, by default it sets suffix values for all elements or indices of the component, rather than the component
itself. Because of this, when we try to retrieve the suffix value for the `model.y`
component, we find that it is `None`.

Suffix values can also be cleared, which is equivalent to setting the value `None`:

```
model.y[3].clear_suffix_value(model.foo)

print(model.y.get_suffix_value(model.foo))    # None
print(model.y[1].get_suffix_value(model.foo)) # 3.0
print(model.y[2].get_suffix_value(model.foo)) # 4.0
print(model.y[3].get_suffix_value(model.foo)) # None
```

4.10 Other Modeling Components

This chapter presented details for some of the most common modeling components
supported by Pyomo. There are other modeling components that were not thor-
oughly discussed in this chapter. These include:

Block: The `Block` component provides a mechanism to declare models with repeated or nested structure (e.g., separate `Block` objects may exist on a model to
represent different time points in a multi-period optimization). A `Block` consists of a collection of Pyomo modeling components. More discussion of blocks
is provided in Chapter 8.

Model: The `Model` component provides a container for grouping Pyomo modeling
components to form the definition of an optimization problem. Pyomo supports
both abstract and concrete modeling representations. While "model" objects
were widely used in this book (they are required to formulate and solve an
optimization problem in Pyomo), we have not discussed the fact that they are
components themselves. In fact, they inherit from the `Block` component.

Complementarity: This component is used to define complementarity conditions in a mathematical program with equilibrium constraints (MPEC). Several forms of the complementarity conditions are supported. This component is documented further in Chapter 13.

ContinuousSet: This component is used to represent bounded continuous domains in the context of modeling differential equations. This component is documented further in Chapter 12.

DerivativeVar: This component is used to represent derivatives of Var components in the context of modeling differential equations. This component is documented further in Chapter 12.

Disjunct: This component supports the Generalized Disjunctive Programming (GDP) capability within Pyomo. A Disjunct component is a container for an indicator variable and a set of constraints that should be active when that indicator variable is True. This component is documented further in Chapter 11.

Disjunction: This component supports the Generalized Disjunctive Programming (GDP) capability within Pyomo. A Disjunction component contains a set of Disjunct objects connected by a logical "OR" operator. This component is documented further in Chapter 11.

Piecewise: This component supports piecewise modeling of general functions. It supports several different transformations to produce mixed-integer representations for the piecewise functions. Additional documentation for this component can be found at the Pyomo website.

SOSConstraint: Special ordered sets (SOS) can be defined in Pyomo through the SOSConstraint component. Pyomo supports special ordered sets of type 1 and 2 (SOS1 and SOS2). Additional documentation for this component can be found at the Pyomo website.

BuildAction and BuildCheck: These components are used mainly in abstract models and are described in Section 10.4.

Chapter 5
Scripting Custom Workflows

Abstract This chapter illustrates the use of Python with Pyomo for solution analysis and the development of custom workflows or high-level meta-algorithms. For example, the chapter shows how to access variable and objective values, add and remove constraints, and iterate over model components. This chapter also contains some larger examples, to illustrate how Pyomo users can go beyond the basics and develop custom solution and analysis strategies.

5.1 Introduction

In previous chapters, we have described how a generic optimization process can be executed with Pyomo to construct a model, solve the model, and display the results. The use of Python and the Pyomo API provides tremendous flexibility for the development of advanced workflows. With some AMLs, a new scripting language is defined that is unique to the AML, and the developers of the package produce a parser for the new language. This separates the user from the underlying code of the framework itself. With Pyomo, Python is used for both the overall framework and the modeling environment. This provides the user with complete control over the entire solution process giving two important high-level capabilities:

- Pyomo users can leverage existing Python libraries for analysis of data before and after solving the optimization problem.
- Pyomo supports development of algorithms requiring problem transformations and multiple solves of problems with different structure and data. Coupled with the programming capabilities of Python, this allows users to build high-level algorithms (e.g., Bender's decomposition, MINLP solvers, and multi-stage initialization strategies).

M. L. Bynum et al., *Pyomo — Optimization Modeling in Python*, Springer Optimization and Its Applications 67, https://doi.org/10.1007/978-3-030-68928-5_5

In this chapter, the basics of scripting with Pyomo will be discussed. This functionality will be demonstrated on some examples, including a Sudoku solver.

> **NOTE:** This chapter shows the power of Pyomo that can be accessed through the Python language, and examples in this chapter may make use of methods on components that are part of the core Pyomo infrastructure. The developers of Pyomo try to maintain backwards compatibility where possible. However, note that the methods described in this chapter are more likely to change than other capabilities discussed in this book.

In Chapter 3 we introduced a warehouse location problem. This problem solved for the optimal locations to build warehouses to meet delivery demands. Please refer back to Section 3.2 for a detailed description of this problem.

The following example, which builds from the previous warehouse location example, highlights the basic pieces found in almost any Python script to solve a problem in Pyomo: (1) load the necessary data, (2) create the Pyomo model, (3) perform optimization of the Pyomo model using a solver interface, and (4) retrieve and report the solution.

```python
import json
import pyomo.environ as pyo
from warehouse_model import create_wl_model

# load the data from a json file
with open('warehouse_data.json', 'r') as fd:
    data = json.load(fd)

# call function to create model
model = create_wl_model(data, P=2)

# solve the model
solver = pyo.SolverFactory('glpk')
solver.solve(model)

# look at the solution
model.y.pprint()
```

This script requires two additional files. The standard Python distribution includes support for reading and writing JSON files. The file below contains the necessary data in JSON format.

```json
{
    "WH": [
        "Harlingen",
        "Memphis",
        "Ashland"
    ],
    "CUST": [
        "NYC",
        "LA",
        "Chicago",
```

```
      "Houston"
    ],
    "dist": {
      "Harlingen": {
        "NYC": 1956,
        "LA": 1606,
        "Chicago": 1410,
        "Houston": 330
      },
      "Memphis": {
        "NYC": 1096,
        "LA": 1792,
        "Chicago": 531,
        "Houston": 567
      },
      "Ashland": {
        "NYC": 485,
        "LA": 2322,
        "Chicago": 324,
        "Houston": 1236
      }
    }
}
```

The warehouse model is defined with the following code.

```
import pyomo.environ as pyo

def create_wl_model(data, P):
    # create the model
    model = pyo.ConcreteModel(name="(WL)")
    model.WH = data['WH']
    model.CUST = data['CUST']
    model.dist = data['dist']
    model.P = P
    model.x = pyo.Var(model.WH, model.CUST, bounds=(0,1))
    model.y = pyo.Var(model.WH, within=pyo.Binary)

    def obj_rule(m):
        return sum(m.dist[w][c]*m.x[w,c] for w in m.WH for c in \
            m.CUST)
    model.obj = pyo.Objective(rule=obj_rule)

    def one_per_cust_rule(m, c):
        return sum(m.x[w,c] for w in m.WH) == 1
    model.one_per_cust = pyo.Constraint(model.CUST, \
        rule=one_per_cust_rule)

    def warehouse_active_rule(m, w, c):
        return m.x[w,c] <= m.y[w]
    model.warehouse_active = pyo.Constraint(model.WH, \
        model.CUST, rule=warehouse_active_rule)

    def num_warehouses_rule(m):
        return sum(m.y[w] for w in m.WH) <= m.P
```

```
    model.num_warehouses = \
        pyo.Constraint(rule=num_warehouses_rule)

    return model
```

Python allows scripting of custom workflows much more powerful than this simple example. In this chapter, a description of the steps to access model components and modify the model programmatically will be illustrated

> **NOTE:** Scripting is possible with abstract models although it is most common to interact with concrete models when scripting. While the examples in this chapter focus on concrete models, a concrete model can be created from an abstract model using `create_instance()`. See Chapter 10 for more details.

5.2 Interrogating the Model

Instances of Pyomo modeling objects (e.g., `Var`, `Param`, `Constraint`) have several attributes that can be accessed programmatically. These attributes can be interrogated to provide information about the state of the model. In this section, we show how to iterate over model components and access key attributes of these components. Chapter 4 provides additional details about the attributes of Pyomo modeling components.

For example, one can access the actual expression for the objective function. Consider the warehouse location problem described previously. The following code will print the expression for the objective function, the value of the objective at the solution, and the value of one of the variables.

```
import json
import pyomo.environ as pyo
from warehouse_model import create_wl_model

# load the data from a json file
with open('warehouse_data.json', 'r') as fd:
    data = json.load(fd)

# call function to create model
model = create_wl_model(data, P=2)

# solve the model
solver = pyo.SolverFactory('glpk')
solver.solve(model)

# print the expression for the objective function
print(model.obj.expr)

# print the value of the objective function
# at the solution
```

```
print(pyo.value(model.obj))

# print the value of a particular variable
print(pyo.value(model.y['Harlingen']))
```

5.2.1 The `value` Function

Some Pyomo component attributes may contain Pyomo objects instead of native Python numerical values. For example, the lower bound of a `Var` may be a simple `float` (e.g., 3.2), but it may also be a Pyomo `Param` object with an associated value. Because of this, it is important to use the `value` function to evaluate attributes that are expected to return numerical values.

The use of `value` is important for converting Pyomo expressions to numeric values. This is illustrated in the following example.

```
import pyomo.environ as pyo

model = pyo.ConcreteModel()
model.u = pyo.Var(initialize=2.0)

# unexpected expression instead of value
a = model.u - 1
print(a) # "u - 1"
print(type(a)) # <class \
    'pyomo.core.expr.numeric_expr.SumExpression'>

# correct way to access the value
b = pyo.value(model.u) - 1
print(b) # 1.0
print(type(b)) # <class 'float'>
```

In this example, a contains a Pyomo expression object (and not a numerical value as might have been expected). This is demonstrated when we print it and check the type. The print statement is correctly showing a description of the expression. To obtain the numeric value of the variable (or other Pyomo components), `pyo.value` needs to be called. Here, b contains the numeric value as expected.

This example illustrates the creation of implicit expressions and how they can lead to unintended consequences. Users are strongly encouraged to avoid generating expressions except as part of the model construction process.

NOTE: It is common to forget the `value()` function when retrieving values from Pyomo `Var` components. For example, in the code shown below, Python will actually print the representation of the variable object itself, not the value.

```
print(model.y)
```

In this case, the code will print the variable name "y", not the value.

5.2.2 Accessing Attributes of Indexed Components

The previous examples illustrate the procedure to access values of scalar variables.
The value of a particular index of an indexed variable by can be retrieved by specifying the exact index.

```
print(pyo.value(model.y['Ashland']))
```

It is also possible to access all the values by iterating over each element of an indexed
variable.

```
for i in model.y:
    print('{0} = {1}'.format(model.y[i], pyo.value(model.y[i])))
```

The loop above iterates over the *keys* for the indexed variable model.y. This example could also have been written to iterate over the index set directly as shown
below.

```
for i in model.WH:
    print('{0} = {1}'.format(model.y[i], pyo.value(model.y[i])))
```

This approach can also be used to access different attributes of variables (e.g.,
lower bound model.y[i].lb) and other model components.

5.2.2.1 Slicing Over Indices of Components

Pyomo supports advanced slicing notation to allow for more control when looping
over individual elements of a model component. For example, we can see all of the
customers that are served from a particular warehouse with the following slicing
notation:

```
for v in model.x['Ashland',:]:
    print('{0} = {1}'.format(v, pyo.value(v)))
```

> **NOTE:** Note that the slicing notation is returning the component objects them-
> selves, and not their indices. Additional information can be found in Sec-
> tion 8.5.

5.2.2.2 Iterating Over All Var Objects on a Model

Pyomo also provides methods to generically loop over components on a given model
or block. In this short example, we show how to iterate over all Var objects on a
model.

```
# loop over the Var objects on the model
for v in model.component_objects(ctype=pyo.Var):
    for index in v:
```

```
    print('{0} <= {1}'.format(v[index], \
        pyo.value(v[index].ub)))

# or use the following to loop over the individual
# indices of each of the Var objects directly
for v in model.component_data_objects(ctype=pyo.Var):
    print('{0} <= {1}'.format(v, pyo.value(v.ub)))
```

NOTE: These methods can be used to find all components on a Pyomo model and iterate over them. This approach also extends to hierarchical models with `Block` components. For a deeper dive into these methods, consult the online documentation.

5.3 Modifying Pyomo Model Structure

A frequent use case is the need to repeatedly solve models with different parameter values or minimal changes to the constraints. Mutable parameters can be used to efficiently solve models with different parameter values. An example showing the use of the mutable `Param P` is shown in Section 3.3.5.

Within Pyomo, it is possible to modify the structure of a model between solves. For example:

- Objectives and constraints can be activated and deactivated without changing the data stored in the model. A deactivated component will be excluded when the Pyomo model is being sent to the solver.
- Variables can be treated as fixed or unfixed (the default).
- Pyomo also allows addition and removal of modeling components. For example, constraints can be added or removed from a model.

The following example illustrates the use of these approaches for modifying model structure:

```
import pyomo.environ as pyo

model = pyo.ConcreteModel()
model.x = pyo.Var(bounds=(0,5))
model.y = pyo.Var(bounds=(0,1))
model.con = pyo.Constraint(expr=model.x + model.y == 1.0)
model.obj = pyo.Objective(expr=model.y-model.x)

# solve the problem
solver = pyo.SolverFactory('glpk')
solver.solve(model)
print(pyo.value(model.x)) # 1.0
print(pyo.value(model.y)) # 0.0

# add a constraint
```

```
model.con2 = pyo.Constraint(expr=4.0*model.x + model.y == 2.0)
solver.solve(model)
print(pyo.value(model.x))  # 0.33
print(pyo.value(model.y))  # 0.66

# deactivate a constraint
model.con.deactivate()
solver.solve(model)
print(pyo.value(model.x))  # 0.5
print(pyo.value(model.y))  # 0.0

# re-activate a constraint
model.con.activate()
solver.solve(model)
print(pyo.value(model.x))  # 0.33
print(pyo.value(model.y))  # 0.66

# delete a constraint
del model.con2
solver.solve(model)
print(pyo.value(model.x))  # 1.0
print(pyo.value(model.y))  # 0.0

# fix a variable
model.x.fix(0.5)
solver.solve(model)
print(pyo.value(model.x))  # 0.5
print(pyo.value(model.y))  # 0.5

# unfix a variable
model.x.unfix()
solver.solve(model)
print(pyo.value(model.x))  # 1.0
print(pyo.value(model.y))  # 0.0
```

5.4 Examples of Common Scripting Tasks

In this section, we will show a few scripting examples that illustrate the capabilities shown previously. More examples can be found in the Pyomo Gallery (see www.pyomo.org).

5.4.1 Warehouse Location Loop and Plotting

The following example formulates the warehouse location problem and solves it repeatedly to find every possible solution. Each time a solution is found, a new cut is added that excludes that solution, and the problem is solved again to find the next solution. This process is repeated until the problem is infeasible, and no more solutions can be found. The `ConstraintList` component is used to contain the list of cuts; each time through the loop a new cut is added to this component.

```python
import json
import pyomo.environ as pyo
from warehouse_model import create_wl_model
import matplotlib.pyplot as plt

# load the data from a json file
with open('warehouse_data.json', 'r') as fd:
    data = json.load(fd)

# call function to create model
model = create_wl_model(data, P=2)
model.integer_cuts = pyo.ConstraintList()
objective_values = list()
done = False
while not done:
    # solve the model
    solver = pyo.SolverFactory('glpk')
    results = solver.solve(model)

    term_cond = results.solver.termination_condition
    print('')
    print('--- Solver Status: {0} ---'.format(term_cond))

    if pyo.check_optimal_termination(results):
        # look at the solution
        print('Optimal Obj. Value = \
            {0}'.format(pyo.value(model.obj)))
        objective_values.append(pyo.value(model.obj))
        model.y.pprint()

        # create new integer cut to exclude this solution
        WH_True = [i for i in model.WH if pyo.value(model.y[i]) > \
            0.5]
        WH_False = [i for i in model.WH if pyo.value(model.y[i]) \
            < 0.5]
        expr1 = sum(model.y[i] for i in WH_True)
        expr2 = sum(model.y[i] for i in WH_False)
        model.integer_cuts.add(
            sum(model.y[i] for i in WH_True) \
            - sum(model.y[i] for i in WH_False) \
            <= len(WH_True)-1)
    else:
        done = True

x = range(1, len(objective_values)+1)
plt.bar(x, objective_values, align='center')
plt.gca().set_xticks(x)
plt.xlabel('Solution Number')
plt.ylabel('Optimal Obj. Value')
plt.savefig('WarehouseCuts.pdf')
```

This example generates console output that shows each of the solutions encountered. It also generates Figure 5.1 with the package `matplotlib` that shows the value of the optimal objective function for each solution obtained.

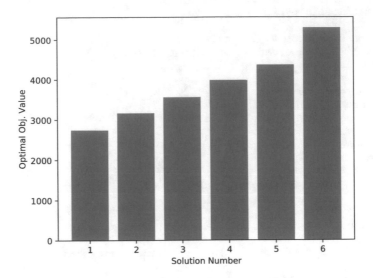

Fig. 5.1: Optimal objective value for a series of solutions obtained from the warehouse location problem.

5.4.2 A Sudoku Solver

In this section, we further illustrate the power of scripting in Python with Pyomo. Specifically, we will solve a feasibility problem and show how to find all the feasible solutions to the Sudoku puzzle. We will solve the problem once, identify a feasible solution, then add an integer cut to remove this solution from the list of possible solutions, and solve the problem again.

A typical Sudoku puzzle is shown in Figure 5.2. In this puzzle, one must fill in the empty cells with the numbers 1 through 9. Each row must have only one occurrence of each number. Likewise, each column must only have one occurrence of each number. Finally, each of the nine sub-squares must also only have one occurrence of each number. We define the sets *ROWS*, *COLS*, and *VALUES* (all of which contain the integers 1 through 9. We then define a binary variable $y[r, c, v]$ to indicate which number is in each of the cells. If $y[r, c, v] = 1$, then this implies that the value v has been selected for the cell identified by row r and column c.

5	3			7				
6			1	9	5			
	9	8					6	
8				6				3
4			8		3			1
7				2				6
	6					2	8	
			4	1	9			5
				8			7	9

Fig. 5.2: An example of a Sudoku puzzle prior to solving.

Using this notation, it is relatively straightforward to define the constraints that restrict the allowable numbers in each row and column as,

$$\sum_{c \in COLS} y[r,c,v] = 1 \quad \forall\, r \in ROWS,\ v \in VALUES$$

$$\sum_{r \in ROWS} y[r,c,v] = 1 \quad \forall\, c \in COLS,\ v \in VALUES$$

The Pyomo code for these constraints is:

```
# exactly one number in each row
def _RowCon(model, r, v):
    return sum(model.y[r,c,v] for c in model.COLS) == 1
model.RowCon = pyo.Constraint(model.ROWS, model.VALUES, \
    rule=_RowCon)

# exactly one number in each column
def _ColCon(model, c, v):
    return sum(model.y[r,c,v] for r in model.ROWS) == 1
model.ColCon = pyo.Constraint(model.COLS, model.VALUES, \
    rule=_ColCon)
```

Defining the constraint that restricts the number for the sub-squares is a little more difficult. To make the definition easier, we define a set with an index for each of the sub-squares. Then, we define a list of tuples that describes the map from each of the sub-squares to the list of corresponding indices. This list, along with the corresponding sub-squares constraint, is defined in the complete code listing for this example at the end of this section. The desired constraint for the sub-squares is given by,

$$\sum_{(r,c)\in ss_{map}[i]} y[r,c,v] = 1 \quad \forall i \in SUBSQUARES.$$

The Pyomo code for this constraint is:

```
# exactly one number in each subsquare
def _SqCon(model, s, v):
    return sum(model.y[r,c,v] for (r,c) in \
        subsq_to_row_col[s]) == 1
model.SqCon = pyo.Constraint(model.SUBSQUARES, model.VALUES, \
    rule=_SqCon)
```

The last key constraint for the Sudoku problem is to make sure that there is only one value allowed per cell. The constraint is given by,

$$\sum_{v\in VALUES} y[r,c,v] = 1 \quad \forall r \in ROWS, c \in COLS.$$

The Pyomo code for this constraint is:

```
# exactly one number in each cell
def _ValueCon(model, r, c):
    return sum(model.y[r,c,v] for v in model.VALUES) == 1
model.ValueCon = pyo.Constraint(model.ROWS, model.COLS, \
    rule=_ValueCon)
```

When designing Sudoku puzzles, two features may change frequently: the initial board layout and the number of integer cuts to remove previously seen solutions. One way to handle this variety of potential inputs is to define a function to create the model from a starting puzzle as well as a list of integer cuts. However, such a function would be inefficient for our purposes since we would be creating an entirely new model each time we wanted to add a single new integer cut after each solve. Thus, we will define two separate functions: one that creates the initial model given a Sudoku board, and another that adds a new integer cut to the given model based on the current value of its variables.

We define an integer cut using two sets. The first set S_0 consists of indices for those variables whose current solution is 0, and the second set S_1 consists of indices for those variables whose current solution is 1. Given these two sets, an integer cut constraint that would prevent such a solution from appearing again is defined by,

$$\sum_{(r,c,v)\in S_0} y[r,c,v] + \sum_{(r,c,v)\in S_1} (1 - y[r,c,v]) \geq 1.$$

The following Python code defines three functions. The first, create_sudoku_model creates the Pyomo model for the Sudoku problem. The second, add_integer_cut creates an integer cut corresponding to the current solution and adds it to the ConstraintList called IntegerCuts. The third, print_solution prints the current solution in the form of a Sudoku board.

```python
import pyomo.environ as pyo

# create a standard python dict for mapping subsquares to
# the list (row,col) entries
subsq_to_row_col = dict()

subsq_to_row_col[1] = [(i,j) for i in range(1,4) for j in range(1,4)]
subsq_to_row_col[2] = [(i,j) for i in range(1,4) for j in range(4,7)]
subsq_to_row_col[3] = [(i,j) for i in range(1,4) for j in range(7,10)]

subsq_to_row_col[4] = [(i,j) for i in range(4,7) for j in range(1,4)]
subsq_to_row_col[5] = [(i,j) for i in range(4,7) for j in range(4,7)]
subsq_to_row_col[6] = [(i,j) for i in range(4,7) for j in range(7,10)]

subsq_to_row_col[7] = [(i,j) for i in range(7,10) for j in range(1,4)]
subsq_to_row_col[8] = [(i,j) for i in range(7,10) for j in range(4,7)]
subsq_to_row_col[9] = [(i,j) for i in range(7,10) for j in range(7,10)]

# creates the sudoku model for a 10x10 board, where the
# input board is a list of fixed numbers specified in
# (row, col, val) tuples.
def create_sudoku_model(board):

    model = pyo.ConcreteModel()

    # store the starting board for the model
    model.board = board

    # create sets for rows columns and squares
    model.ROWS = pyo.RangeSet(1,9)
    model.COLS = pyo.RangeSet(1,9)
    model.SUBSQUARES = pyo.RangeSet(1,9)
    model.VALUES = pyo.RangeSet(1,9)

    # create the binary variables to define the values
    model.y = pyo.Var(model.ROWS, model.COLS, model.VALUES, within=pyo.Binary)

    # fix variables based on the current board
    for (r,c,v) in board:
        model.y[r,c,v].fix(1)

    # create the objective - this is a feasibility problem
    # so we just make it a constant
    model.obj = pyo.Objective(expr= 1.0)

    # exactly one number in each row
    def _RowCon(model, r, v):
        return sum(model.y[r,c,v] for c in model.COLS) == 1
    model.RowCon = pyo.Constraint(model.ROWS, model.VALUES, rule=_RowCon)

    # exactly one number in each column
    def _ColCon(model, c, v):
        return sum(model.y[r,c,v] for r in model.ROWS) == 1
    model.ColCon = pyo.Constraint(model.COLS, model.VALUES, rule=_ColCon)

    # exactly one number in each subsquare
    def _SqCon(model, s, v):
        return sum(model.y[r,c,v] for (r,c) in subsq_to_row_col[s]) == 1
    model.SqCon = pyo.Constraint(model.SUBSQUARES, model.VALUES, rule=_SqCon)

    # exactly one number in each cell
    def _ValueCon(model, r, c):
        return sum(model.y[r,c,v] for v in model.VALUES) == 1
    model.ValueCon = pyo.Constraint(model.ROWS, model.COLS, rule=_ValueCon)

    return model
```

```
# use this function to add a new integer cut to the model.
def add_integer_cut(model):
    # add the ConstraintList to store the IntegerCuts if
    # it does not already exist
    if not hasattr(model, "IntegerCuts"):
        model.IntegerCuts = pyo.ConstraintList()

    # add the integer cut corresponding to the current
    # solution in the model
    cut_expr = 0.0
    for r in model.ROWS:
        for c in model.COLS:
            for v in model.VALUES:
                if not model.y[r,c,v].fixed:
                    # check if the binary variable is on or off
                    # note, it may not be exactly 1
                    if pyo.value(model.y[r,c,v]) >= 0.5:
                        cut_expr += (1.0 - model.y[r,c,v])
                    else:
                        cut_expr += model.y[r,c,v]
    model.IntegerCuts.add(cut_expr >= 1)

# prints the current solution stored in the model
def print_solution(model):
    for r in model.ROWS:
        print(' '.join(str(v) for c in model.COLS
                        for v in model.VALUES
                        if pyo.value(model.y[r,c,v]) >= 0.5))
```

The following code shows a script that drives the optimization process based on
these three functions. This script defines the candidate board, and iteratively solves
the Sudoku problems by adding integer cuts until the problem is no longer feasible.
Infeasibility is assumed when the solver termination condition is no longer reported
as optimal.

```
from pyomo.opt import (SolverFactory,
                       TerminationCondition)
from sudoku import (create_sudoku_model,
                    print_solution,
                    add_integer_cut)

# define the board
board = [(1,1,5),(1,2,3),(1,5,7), \
         (2,1,6),(2,4,1),(2,5,9),(2,6,5), \
         (3,2,9),(3,3,8),(3,8,6), \
         (4,1,8),(4,5,6),(4,9,3), \
         (5,1,4),(5,4,8),(5,6,3),(5,9,1), \
         (6,1,7),(6,5,2),(6,9,6), \
         (7,2,6),(7,7,2),(7,8,8), \
         (8,4,4),(8,5,1),(8,6,9),(8,9,5), \
         (9,5,8),(9,8,7),(9,9,9)]

model = create_sudoku_model(board)

solution_count = 0
while 1:

    with SolverFactory("glpk") as opt:
        results = opt.solve(model)
```

```
    if results.solver.termination_condition != \
       TerminationCondition.optimal:
        print("All board solutions have been found")
        break

solution_count += 1

add_integer_cut(model)

print("Solution #%d" % (solution_count))
print_solution(model)
```

Running this script provides all possible solutions as the output. In this example, there is only one solution to the candidate Sudoku puzzle, as shown in Figure 5.3.

5	3	4	6	7	8	9	1	2
6	7	2	1	9	5	3	4	8
1	9	8	3	4	2	5	6	7
8	5	9	7	6	1	4	2	3
4	2	6	8	5	3	7	9	1
7	1	3	9	2	4	8	5	6
9	6	1	5	3	7	2	8	4
2	8	7	4	1	9	6	3	5
3	4	5	2	8	6	1	7	9

Fig. 5.3: Solved Sudoku puzzle.

Chapter 6
Interacting with Solvers

Abstract This chapter describes how to interface with solvers in Pyomo. Basic functionality supported by all interfaces includes translation of a Pyomo instance into the format required by a solver, solver options processing, solver invocation, solve status checking, and solution loading.

6.1 Introduction

Figure 6.1 shows a high-level view of the relationship between Pyomo and optimization software that we will refer to as a *solver*. Pyomo is a modeling tool and does not, itself, include solvers. Rather Pyomo has several interfaces to solvers. As shown in the figure, Pyomo translates the model into a format used by a solver. The results of the solver are used to populate the `Var` objects in the Pyomo model and a `Results` object is also returned that can be queried to obtain more information about solver execution. Additional information can be sent to the solver (e.g., options) and received from the solver (e.g. dual values) but these are not shown in the figure.

Pyomo models can be analyzed with a wide variety of optimization solvers, and there are several types of solver interfaces in Pyomo:

- A *shell solver* is launched as a separate sub-process by running an executable found on the user's PATH environment. Pyomo interfaces with these solvers through files. Pyomo generates a file description of the problem, launches the solver, and then loads the results from log files and standard output files. This is a common form of solver.
- A *direct solver* is executed as a subroutine. Pyomo interfaces with these solvers through libraries installed and exposed in the form of Python packages. This is a less common form of solver, since it relies on the existence of Python interfaces to solver libraries.
- A *persistent solver* is related to a direct solver, but includes additional capabilities to allow for incremental modification and re-solve of a model. Persistent

M. L. Bynum et al., *Pyomo — Optimization Modeling in Python*, Springer Optimization and Its Applications 67, https://doi.org/10.1007/978-3-030-68928-5_6

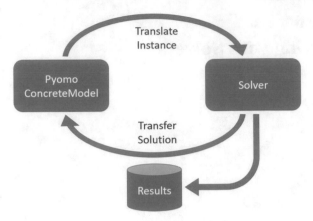

Fig. 6.1: Stylized representation of typical interactions between Pyomo, which processes the model and a solver, which computes solutions.

solvers are typically used to improve performance in situations where model construction time is significant or known solutions to a related family of models can be used to reduce solve times. Persistent solvers are discussed further in Section 9.

The rest of this chapter discusses how to call a solver, send solver-specific options, and obtain solver execution information and results.

6.2 Using Solvers

As seen in Section 5.1 the `SolverFactory` function is used to construct a solver interface object. The argument passed to the solver factory specifies the name of the solver being used. In most cases, this is the name of the executable that will be used to solve the problem; however, Pyomo supports shorter names for some solvers. For example, the GLPK solver can be specified with

```
solver = pyo.SolverFactory('glpk')
```

Once a solver object has been constructed, the solver can be invoked by calling the `solve()` method. The `solve()` method accepts a number of keyword arguments, a few of which are shown here, more or less in order of importance.

- `options`: A dictionary of options to be passed to the underlying solver.
- `tee`: If this argument is `True`, then the solver output is printed both to the standard output as well as saved to the log file. If `False` (the default), then the solver output is only saved to log file if the solver creates one.
- `load_solutions`: If this argument is `True` (the default), then solution values are automatically transfered to `Var` objects on the model. If `False`,

then the results object keeps a *raw* representation of the solutions and it is not transferred to the model. It can be transfered to the model using the `model.solutions.load_from()` method.

- `logfile`: The filename used to store output for shell solvers.
- `solnfile`: The filename used to store the solution for shell solvers.
- `timelimit`: The number of seconds that a shell solver is run before it is terminated. (default is `None`)
- `report_timing`: If this argument is `True`, then timing information is reported by the solver (default is `False`)
- `solver_io`: Specifies an alternative solver interface, e.g. `solver_io='nl'`.
- `suffixes`: A list of suffixes that are exported to the solver.

The `options` attribute can be used to send solver specific options to the underlying solver. In the following example, we pass the `tee=True` keyword argument to tell Pyomo to print the solver's execution trace to the terminal. We pass two solver-specific options to GLPK (sending log output to `warehouse.log` and turning off scaling). Notice that some solver-specific options do not take values (e.g. `noscale`), but are simply flags to turn on or off particular behavior. For these types of options, set the option value in the dictionary to `None`. Note that options can also be sent directly to the `solve` function as a dictionary. Options that are passed to the `solve` function do not persist.

```python
import json
import pyomo.environ as pyo
from warehouse_model import create_wl_model

# load the data from a json file
with open('warehouse_data.json', 'r') as fd:
    data = json.load(fd)

# call function to create model
model = create_wl_model(data, P=2)

# create the solver
solver = pyo.SolverFactory('glpk')

# options can be set directly on the solver
solver.options['noscale'] = None
solver.options['log'] = 'warehouse.log'
solver.solve(model, tee=True)
model.y.pprint()

# options can also be passed via the solve command
myoptions = dict()
myoptions['noscale'] = None
myoptions['log'] = 'warehouse.log'
solver.solve(model, options=myoptions, tee=True)
model.y.pprint()
```

6.3 Investigating the Solution

After a model is solved, there are two aspects of the solution to be investigated. The first is the solution status returned from the solver, and the second is the value of the variables and objective function. The solution status can usually be viewed in the console by passing `tee=True` into the solve command, but there are many times when one would like to read this status in code. This section will discuss the results object, and show how to retrieve values from variables after the solve.

6.3.1 Solver Results

The `solve()` method returns a results object that contains status information from the solver. If the solve completes successfully, the solution values are loaded directly into the model. This consists of three steps: (1) storing solutions in the `solutions` attribute of the model, (2) load the values of variables from a selected solution, and (3) remove solutions from the results object. Afterwards, the `results` object only contains meta-data about the model and the optimization process. For memory efficiency, it no longer contains the solution itself by default.

> **NOTE:** By default, when the solver completes, the solution is automatically loaded into the model object and removed from the results object. Due to this, the results object will indicate it has 0 solutions.

Typically, a solver only returns a single solution; however, there are cases where a solver might return multiple solutions (a pool of solutions). Due to this, the results object supports an interface that looks like a dictionary of lists containing more than one solution. However, for the most common case of a single solution, the results object supports a simple attribute-like interface. The results object returned from the `solve()` method contains a `problem` attribute and a `solver` attribute that contain information about the problem statistics and the solver status.

The `results.solver` attribute contains a `SolverInformation` object. Key attributes of this object are shown in the Table 6.1.

As noted in Section 2.5, the simplest thing to do with the `results` object is to pass it to the function `assert_optimal_termination` , which halts the script and outputs a message if the solver does not report that it found an optimal solution. In situations where the script should not halt if the solution is optimal, but a test for optimality is needed, the `results` object can be passed to the `check_optimal_termination` function that returns `True` if the solver report optimality and `False` if not.

In some scripts, it makes sense to defer moving the solution from the results object to the model until after optimality has been checked. This might be needed for efficieny or to avoid having `Var` values that are not optimal loaded into the

Table 6.1: Key attributes of the `SolverInformation` object

Attribute	
status	Returns the solver status as a member of the `SolverStatus` enum that can be: ok, warning, error, aborted, or unknown.
termination_condition	Returns the specific termination condition as reported by the solver. This is an enum called `TerminationCondition` that can have different values, including optimal, infeasible, or unbounded. There are many different solver outcomes, and depending on the solver, other outcomes may be seen.
termination_message	String message returned by the solver summarizing the termination status.

model. To avoid automatic loading of the solution from the `results` object to the model, use the `load_solutions=False` argument to the call to `solve()`. To move the solution values from the `results` object to the `Var` values in the model, use `model.solutions.load_from(results)`, which uses the `solutions` object that is automatically attached to the `model` object when it is passed to the `solve` method. A short example of these steps is shown:

```
from pyomo.opt import SolverStatus, TerminationCondition
# Wait to load the solution into the model until
# after the solver status is checked
results = solver.solve(model, load_solutions=False)
if (results.solver.status == SolverStatus.ok) and \
    (results.solver.termination_condition == \
    TerminationCondition.optimal):
    # Manually load the solution into the model
    model.solutions.load_from(results)
else:
    print("Solve failed.")
```

Part II
Advanced Topics

Chapter 7
Nonlinear Programming with Pyomo

Abstract This chapter describes the nonlinear programming capabilities of Pyomo. It presents the nonlinear expressions and functions supported, and it provides some tips for formulating and solving nonlinear programming problems. This chapter also provides several real-world examples to illustrate formulating and solving nonlinear programming problems. Finally, it provides a brief discussion of supported solvers for nonlinear problems.

7.1 Introduction

It is not possible to adequately represent many applications without modeling nonlinear relationships. Fortunately, Pyomo has the ability to represent general nonlinear programming (NLP) problems in a straightforward manner. However, the solution of this class of problems presents several challenges that do not exist for linear problems. For example, most modern, efficient NLP solvers require derivatives of the constraints and the objective function. Since the functions are nonlinear, this requires accurate numerical evaluation of these derivatives. Additionally, in the case of non-convex problems, multiple local minima may exist due to the shape of the objective function or the constraints, and specifying a suitable starting point may be critical.

In Section 7.2, we describe the nonlinear expressions supported in Pyomo and then illustrate how to build a nonlinear problem formulation within Pyomo. In Section 7.3, we briefly discuss the solvers supported by Pyomo, and we provide a few tips to help effectively formulate nonlinear programming problems. Finally, we close this chapter with a number of small, but real-world nonlinear programming examples.

7.2 Nonlinear Progamming Problems in Pyomo

Pyomo supports the following general nonlinear programming formulation:

$$\min_{x} \; f(x)$$
$$\text{s.t.} \; c(x) = 0$$
$$d^L \le d(x) \le d^U$$
$$x^L \le x \le x^U.$$

The allowable form of the objective function $f(x)$, the vector of equality constraints $c(x)$, and the vector of inequality constraints $d(x)$ depends entirely on the solver selected to provide a solution. However, Pyomo is tested extensively with local and global solvers that typically assume that these functions are continuous and smooth, with continuous first (and possibly second) derivatives. The development of nonlinear extensions for Pyomo has focused on this broad problem class.

7.2.1 Nonlinear Expressions

Formulating nonlinear optimization problems in Pyomo is no different from formulating linear or mixed-integer problems. All the Pyomo modeling components described throughout the book are used in the same way (e.g., Objective, Constraint) except they may include nonlinear expressions.

Table 7.1 lists the operators currently supported to formulate expressions, with examples where x and y are Pyomo Var objects. In addition to these operators, Pyomo supports a number of nonlinear functions as described in Table 7.2. Note that these are Pyomo-specific functions Pyomo expects, and that nonlinear functions from other Python libraries are *not* supported within Pyomo expressions.

NOTE: Passing Pyomo components (e.g., Var, Param) to nonlinear functions from other Python packages (e.g., math, or numpy) when creating a Pyomo expression will cause an exception to be raised which may be difficult to debug. This mistake can be avoided by always using named imports and by being explicit about where a nonlinear function is coming from.

```
import pyomo.environ as pyo    # Use pyo.sin in Pyomo
import math as mt              # expressions not mt.sin
```

Table 7.1: Python operators that have been redefined to generate Pyomo expressions.

Operation	Operator	Example
multiplication	*	expr = model.x * model.y
division	/	expr = model.x / model.y
exponentiation	**	expr = (model.x+2.0)**model.y
in-place multiplication[1]	*=	expr *= model.x
in-place division[2]	/=	expr /= model.x
in-place exponentiation[3]	**=	expr **= model.x

[1] This example for in-place multiplication is equivalent to expr = expr * model.x.
[2] This example for in-place division is equivalent to expr = expr / model.x.
[3] This example for in-place exponentiation is equivalent to expr = expr ** model.x.

Table 7.2: Functions supported by Pyomo for the definition of nonlinear expressions. This table assumes that Pyomo has been imported with import pyomo.environ as pyo.

Operation	Function	Example
arccosine	acos	expr = pyo.acos(model.x)
hyperbolic arccosine	acosh	expr = pyo.acosh(model.x)
arcsine	asin	expr = pyo.asin(model.x)
hyperbolic arcsine	asinh	expr = pyo.asinh(model.x)
arctangent	atan	expr = pyo.atan(model.x)
hyperbolic arctangent	atanh	expr = pyo.atanh(model.x)
cosine	cos	expr = pyo.cos(model.x)
hyperbolic cosine	cosh	expr = pyo.cosh(model.x)
exponential	exp	expr = pyo.exp(model.x)
natural log	log	expr = pyo.log(model.x)
log base 10	log10	expr = pyo.log10(model.x)
sine	sin	expr = pyo.sin(model.x)
square root	sqrt	expr = pyo.sqrt(model.x)
hyperbolic sine	sinh	expr = pyo.sinh(model.x)
tangent	tan	expr = pyo.tan(model.x)
hyperbolic tangent	tanh	expr = pyo.tanh(model.x)

7.2.2 The Rosenbrock Problem

In this section we present a short example to illustrate the formulation and solution of a nonlinear Pyomo model. We consider the unconstrained minimization of the two-variable Rosenbrock function, which is a classic problem frequently used as an example for discussion of unconstrained nonlinear optimization algorithms (e.g., see [45]). This problem is defined as

$$\min_{x,y} \; f(x,y) = (1-x)^2 + 100\left(y-x^2\right)^2,$$

and the solution is in the bottom of the banana shaped valley at the point $x=1$ and $y=1$ (See Figure 7.1).

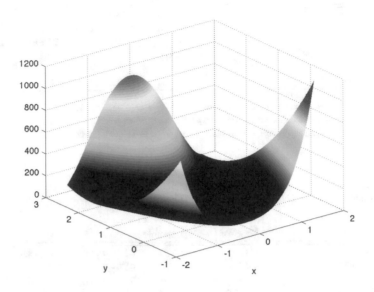

Fig. 7.1: Surface plot of the Rosenbrock function $f(x,y)=(1-x)^2+100\left(y-x^2\right)^2$. The minimum is in the bottom of a banana shaped valley at the point $x=1$, $y=1$.

Consider the following Pyomo model for this problem:

```
# rosenbrock.py
# A Pyomo model for the Rosenbrock problem
import pyomo.environ as pyo

model = pyo.ConcreteModel()
model.x = pyo.Var(initialize=1.5)
model.y = pyo.Var(initialize=1.5)

def rosenbrock(model):
    return (1.0 - model.x)**2 \
        + 100.0*(model.y - model.x**2)**2
model.obj = pyo.Objective(rule=rosenbrock, sense=pyo.minimize)

status = pyo.SolverFactory('ipopt').solve(model)
pyo.assert_optimal_termination(status)
model.pprint()
```

This example illustrates that defining a nonlinear model is really no different from defining a linear model. The model creates two variables x and y and initializes each of them to a value of 1.5. Notice that there is no need to provide any indication the variables will later appear in a nonlinear expression; this will be deduced by Pyomo before solving the problem. The construction rule for the objective function simply returns a nonlinear expression. The nonlinear solver IPOPT is then used to solve the problem, followed by a check of the solver status and printing of the solution.

> **NOTE:** *Pyomo should work with any AMPL-based or GAMS-based solver.* Thus a number of competitive, commercial and open-source packages can be used to solve Pyomo models. For the examples in this chapter, IPOPT [34], an open-source nonlinear optimization package, is used.

The Rosenbrock example can be executed with the command:

```
python rosenbrock.py
```

This produces output similar to the following:

```
2 Var Declarations
    x : Size=1, Index=None
        Key   : Lower : Value                    : Upper : Fixed :
    Stale : Domain
        None :  None : 1.0000000000008233 :  None : False :
    False :  Reals
    y : Size=1, Index=None
        Key   : Lower : Value                    : Upper : Fixed :
    Stale : Domain
        None :  None : 1.0000000000016314 :  None : False :
    False :  Reals

1 Objective Declarations
    obj : Size=1, Index=None, Active=True
        Key   : Active : Sense     : Expression
        None :   True : minimize : (1.0 - x)**2 + 100.0*(y -
    x**2)**2

3 Declarations: x y obj
```

In this output, we see that the problem is correctly solved to a value of $x=y=1.0$, with an objective value of essentially zero. While this example has only a single nonlinear objective and two scalar variables, the modeling components discussed in earlier chapters may also be used.

7.3 Solving Nonlinear Programming Formulations

In most cases, an appropriate nonlinear solver must be installed before Pyomo can be used to optimize a nonlinear programming model. Pyomo's capabilities are focused on modeling optimization applications, and there are a limited number of solvers directly integrated with Pyomo.

7.3.1 Nonlinear Solvers

Nonlinear programming solvers require the modeling framework to evaluate the objective function and constraints at candidate points in x. As well, many nonlinear solvers also require evaluation of first, and often second, derivatives at candidate points. However, a Pyomo user does not have to implement these computations. Instead, automatic differentiation (AD) tools are used to provide accurate and efficient numerical evaluation of the first and second derivatives without any user involvement.

Pyomo provides interfaces for nonlinear solvers compiled for either AMPL or GAMS, and it utilizes the model evaluation and automatic differentiation (AD) capabilities provided by these tools to support efficient computation of derivatives. By supporting these interfaces, a wide array of solvers are immediately available for use with Pyomo without the need to develop individual interfaces for each solver.

In the case of AMPL solvers, the solver code itself is compiled with the AMPL Solver Library (ASL) interface [23], which is in the public domain. Therefore, there are a number of open-source solvers available to Pyomo through this interface. The Pyomo-GAMS interface performs a translation of the Pyomo model into a GAMS model, and consequently a user will need to install GAMS to use GAMS-specific solvers.

7.3.2 Additional Tips for Nonlinear Programming

Effective formulation and solution of nonlinear programming problems can be significantly more challenging than linear programming problems. In this section, we provide a few basic tips to help with the formulation and solution of nonlinear programming problems.

Variable Initialization

Solvers for nonlinear programming problems often require the initialization of problem variables. If initial values are not specified, then Pyomo assumes that the initial values are zero. However, these default values cannot be relied on in many applications.

For the general nonconvex case, nonlinear programming problems can, and often do, have multiple local solutions. While significant advances have been made in global optimization (i.e., rigorous methods that provide a guarantee of global optimality), general large-scale problems are often still intractable, even with state-of-the-art global solvers. Consequently, one is often forced to employ a solver that only provides a guarantee of local optimality. And, it is often critical to initialize the problem effectively to ensure convergence to a desirable local solution.

Sometimes, the undesired local solutions are not physically meaningful, and a sensible initialization with reasonable variable bounds is sufficient to ensure reliable progress to the desired solution. Other times, there may be several physically reasonable local solutions. The development of good nonlinear problem formulations often includes significant effort to provide a reasonable initialization strategy.

Undefined Evaluations

Several nonlinear functions are only well defined over a specific domain (e.g., $log(x)$ is only valid for $x > 0$). Therefore, the modeler must take care to ensure the problem formulation restricts the variable values to be within a valid domain. This is usually accomplished by setting reasonable bounds and initial values on the variables.

It is also important to note that many nonlinear solvers use first (and sometimes second) derivative information for the objective function and the constraints. Therefore, one may need to also restrict the variables to be within a valid domain for the derivatives of the nonlinear expressions. For example, when `sqrt(x)` is included in an expression, then specifying the bounds $x \geq 0$ may not be sufficient. While $\sqrt{x}$ is valid at $x=0$, its derivative, $1/\sqrt{x}$ is not. This should be considered when setting reasonable variable bounds.

Finally, note that some nonlinear interior-point solvers (e.g., IPOPT) may relax the variable bounds slightly before solving the problem. While this has proven to be an effective strategy in many applications, this can sometimes cause a domain violation even if the modeler has specified reasonable variable bounds. One may need to disable this behavior in the solver or apply more conservative bounds.

Model Singularities and Problem Scaling

Many nonlinear programming solvers have restrictions on the constraints (called constraint qualifications) that must be satisfied to guarantee convergence. In particular, it is often a good idea to ensure the constraints are independent everywhere within the solution domain (i.e., the set of active constraint gradients are linearly independent). Nocedal and Wright [45] discuss this issue further (see Chapter 12).

Unfortunately, a model satisfying these restrictions in exact math may still exhibit problems when solved numerically. If the model is ill-conditioned, then many solvers can have difficulty converging or finding a solution efficiently. It is important to scale the model to provide a well-conditioned Jacobian and Hessian. This can be as simple as linearly scaling the variables and the constraints. In difficult cases, the model may need to be reformulated.

7.4 Nonlinear Programming Examples

In this section we present several examples that illustrate the capabilities of Pyomo with nonlinear problems.

7.4.1 Variable Initialization for a Multimodal Function

The following example illustrates the importance of effective variable initialization. Consider the minimization of the following multimodal function:

$$f(x) = (2 - \cos(\pi x) - \cos(\pi y))\, x^2 y^2,$$

which has multiple local minima. The following Pyomo model for this problem initializes the variables at $x=y=0.25$.

```python
# multimodal_init1.py
import pyomo.environ as pyo
from math import pi

model = pyo.ConcreteModel()
model.x = pyo.Var(initialize = 0.25, bounds=(0,4))
model.y = pyo.Var(initialize = 0.25, bounds=(0,4))

def multimodal(m):
    return (2-pyo.cos(pi*m.x)-pyo.cos(pi*m.y)) * (m.x**2) * \
        (m.y**2)
model.obj = pyo.Objective(rule=multimodal, sense=pyo.minimize)

status = pyo.SolverFactory('ipopt').solve(model)
pyo.assert_optimal_termination(status)
print(pyo.value(model.x), pyo.value(model.y))
```

This problem can be solved with the command:

```
python multimodel_init1.py
```

IPOPT finds the solution close to our initial point $x=y=0.0178$. However, if we change the problem and initialize the variables at $x=y=2.1$,

```python
model.x = pyo.Var(initialize = 2.1, bounds=(0,4))
model.y = pyo.Var(initialize = 2.1, bounds=(0,4))
```

then IPOPT finds a different local solution at $x=y=2.0$.

7.4.2 Optimal Quotas for Sustainable Harvesting of Deer

Maintaining a healthy deer population relies on both the management of effective habitats and a sustainable harvesting policy. Among most hunters there is high demand for tags that allow them to take bucks. However, harvesting too many bucks within a population can limit future population growth. We consider a nonlinear programming formulation that determines an optimal policy for deer harvesting by maximizing the value of the harvest while maintaining a strong and sustainable deer population.

Consider a model adapted from Bailey [5] describing the dynamics of the deer population. The deer population in a given area can be divided into three sub-populations: bucks, does, and fawns. Additionally, each year is divided into four periods: winter, breeding season, summer, and harvest. The model describing the population dynamics is based on the following assumptions:

- It is assumed the sub-populations can be represented by continuous variables (i.e., population numbers are large enough to make this a good approximation).
- Each season, there is a reduction in the number of bucks, does, and fawns. This reduction is assumed to be due to natural causes and is proportional to the size of the sub-populations. This reduction is captured by specifying a fractional survival rate depending on the period (winter, breeding, summer, harvest) and the sub-population in question (bucks, does, fawns).
- New fawns are born each year during the breeding season. Fawns are born from does and older fawns according to a birth rate depending on the available amount of food. Half of them are assumed to be male and half are assumed to be female. After surviving one year, half of the remaining fawns become bucks and half become does.
- The total yearly food supply is constant and represents a constraint based on habitat management.
- All harvesting is based on hunting. Hunting quotas can be set for each sub-population, and these quotas are assumed to be completely filled (i.e., all hunters are successful).

The complete derivation of the sub-population model is given in [5], resulting in the following set of difference equations,

$$f_{y+1} = p_1 br_y \left(\frac{p_2}{10} f_y + p_3 d_y \right) - h_y^f \tag{7.1}$$

$$d_{y+1} = p_4 d_y + \frac{p_5}{2} f_y - h_y^d \tag{7.2}$$

$$b_{y+1} = p_6 b_y + \frac{p_5}{2} f_y - h_y^b \tag{7.3}$$

$$br_y = 1.1 + 0.8 \frac{p_s - c_y}{p_s} \tag{7.4}$$

$$c_y = p_7 b_y + p_8 d_y + p_9 f_y \tag{7.5}$$

where the value for parameters p_1 through p_9 are calculated from the various survival rates and food consumption rates. These values are given in Table 7.3. The variables f_y, d_y, and b_y represent the number of fawns, does, and bucks in year y, respectively. Likewise, h_y^f, h_y^d, and h_y^b are the unknown numbers of fawns, does, and bucks harvested in year y, respectively. The birth rate br_y for does is described by a nonlinear relationship where c_y is the amount of food consumed by the deer (in pounds) and p_s is the total available supply of food (again in pounds).

parameter	value	parameter	value
p_1	0.88	p_7	2700.0
p_2	0.82	p_8	2300.0
p_3	0.92	p_9	540.0
p_4	0.84	w_f	1.0
p_5	0.73	w_d	1.0
p_6	0.87	w_b	10.0
p_s	700000		

Table 7.3: Parameter values used by the deer harvesting problem.

In the original reference, this set of difference equations was optimized in the formulation over a period of 20 years so that a sustainable steady-state policy could be deduced from the values at later years. Here, instead include only one year and add the constraint that the number of fawns, does, and bucks at year $y+1$ is equal to those at y. This provides the same steady-state solution with a significantly smaller formulation.

The objective is to maximize the value of the harvest, giving the following nonlinear programming formulation,

$$\max \ w_b h_y^b + w_f h_y^f + w_d h_y^d \tag{7.6}$$

$$f_y = p_1 br_y \left(\frac{p_2}{10} f_y + p_3 d_y \right) - h_y^f \tag{7.7}$$

$$d_y = p_4 d_y + \frac{p_5}{2} f_y - h_y^d \tag{7.8}$$

$$b_y = p_6 b_y + \frac{p_5}{2} f_y - h_y^b \tag{7.9}$$

$$br_y = 1.1 + 0.8 \frac{p_s - c_y}{p_s} \tag{7.10}$$

$$c_y = p_7 b_y + p_8 d_y + p_9 f_y \tag{7.11}$$

$$c_y \leq p_s \tag{7.12}$$

$$b_y \geq \frac{1}{5}(0.4 f_y + d_y) \tag{7.13}$$

where w_f, w_d and w_b represent the value of harvesting a fawn, doe, and buck, respectively. As can be seen in Table 7.3, it is assumed that the value of a buck tag is 10 times the value of a doe or fawn tag. Equation (7.12) ensures that the amount of consumed food cannot be more than the available supply, thereby restricting the overall size of the population. Equation (7.13) ensures that the number of bucks is large enough for effective, sustainable breeding.

The following Pyomo model represents the optimal deer harvesting problem:

```python
# DeerProblem.py
import pyomo.environ as pyo

model = pyo.AbstractModel()

model.p1 = pyo.Param();
model.p2 = pyo.Param();
model.p3 = pyo.Param();
model.p4 = pyo.Param();
model.p5 = pyo.Param();
model.p6 = pyo.Param();
model.p7 = pyo.Param();
model.p8 = pyo.Param();
model.p9 = pyo.Param();
model.ps = pyo.Param();

model.f = pyo.Var(initialize = 20, within=pyo.PositiveReals)
model.d = pyo.Var(initialize = 20, within=pyo.PositiveReals)
model.b = pyo.Var(initialize = 20, within=pyo.PositiveReals)

model.hf = pyo.Var(initialize = 20, within=pyo.PositiveReals)
model.hd = pyo.Var(initialize = 20, within=pyo.PositiveReals)
model.hb = pyo.Var(initialize = 20, within=pyo.PositiveReals)

model.br = pyo.Var(initialize=1.5, within=pyo.PositiveReals)

model.c = pyo.Var(initialize=500000, within=pyo.PositiveReals)


def obj_rule(m):
    return 10*m.hb + m.hd + m.hf
model.obj = pyo.Objective(rule=obj_rule, sense=pyo.maximize)

def f_bal_rule(m):
    return m.f == m.p1*m.br*(m.p2/10.0*m.f + m.p3*m.d) - m.hf
model.f_bal = pyo.Constraint(rule=f_bal_rule)

def d_bal_rule(m):
    return m.d == m.p4*m.d + m.p5/2.0*m.f - m.hd
model.d_bal = pyo.Constraint(rule=d_bal_rule)

def b_bal_rule(m):
    return m.b == m.p6*m.b + m.p5/2.0*m.f - m.hb
model.b_bal = pyo.Constraint(rule=b_bal_rule)

def food_cons_rule(m):
    return m.c == m.p7*m.b + m.p8*m.d + m.p9*m.f
model.food_cons = pyo.Constraint(rule=food_cons_rule)
```

```python
def supply_rule(m):
    return m.c <= m.ps
model.supply = pyo.Constraint(rule=supply_rule)

def birth_rule(m):
    return m.br == 1.1 + 0.8*(m.ps - m.c)/m.ps
model.birth = pyo.Constraint(rule=birth_rule)

def minbuck_rule(m):
    return m.b >= 1.0/5.0*(0.4*m.f + m.d)
model.minbuck = pyo.Constraint(rule=minbuck_rule)

# create the ConcreteModel
instance = model.create_instance('DeerProblem.dat')
status = pyo.SolverFactory('ipopt').solve(instance)
pyo.assert_optimal_termination(status)

instance.pprint()
```

The following data file represents the parameters in Table 7.3:

```
# DeerProblem.dat
param p1 := 0.88;
param p2 := 0.82;
param p3 := 0.92;
param p4 := 0.84;
param p5 := 0.73;
param p6 := 0.87;
param p7 := 2700;
param p8 := 2300;
param p9 := 540;
param ps := 700000;
```

This problem can be optimized with the command:

```
python DeerProblem.py
```

The optimal solution is to harvest 62 bucks, 37 does, and no fawns. This solution favors harvesting of bucks, but harvesting too many bucks would affect population growth. The residual for the `minbuck` constraint is essentially zero, which means that this inequality constraint is binding at the solution. Therefore, this constraint is restricting the number of bucks that can be harvested.

Obviously, this solution is a function of the parameter values that determine the value of fawns, does, and bucks in the objective function, as well as the parameters in model for the population dynamics. Because Pyomo is built with Python, it is straightforward to develop a script to determine the optimal solution as a function of different parameter values, enabling more advanced analysis of the system. Chapter 5 gives more discussion of this functionality.

7.4.3 Estimation of Infectious Disease Models

Effective widespread vaccination programs have significantly minimized the impact of many childhood diseases. However, childhood infectious diseases continue to be a concern in developing countries, and outbreaks of new disease strains pose challenges for public health policy makers. In this example, we simulate the outbreak of an infectious disease within a small community of 300 individuals (representing, for example, a small school). We derive a basic model to describe the spread of infection in the population and use a nonlinear programming formulation to estimate key parameters in this model using the simulated data.

We use a standard discrete time compartment model to represent the system. Individuals are separated into three compartments based on their status with respect to the disease: susceptible (S), infected (I), or recovered (R). We assume that once an individual has contracted the disease and recovered, they are immune from that point forward (i.e., they do not return to the susceptible pool). The discrete time model representing this system is given by:

$$I_i = \frac{\beta I_{i-1}^{\alpha} S_{i-1}}{N}$$

$$S_i = S_{i-1} - I_i$$

These two difference equations describe the propagation of the disease in the population. As a generation-based model, it is assumed that all the individuals infected at time i have recovered by time $i + 1$. I_i and S_i are the number of infected and susceptible individuals at time i, respectively. The population size is given by N, and β and α are model parameters.

> **NOTE:** Typically, we refer to *parameters* as fixed data in our optimization problem. However, in this example, the parameters in our infectious disease model are not yet known, and we want to estimate them from existing data. As a result, model parameters β and α become Pyomo variables in the model (since they are to be estimated with the optimization).

In this example, we use least-squares to estimate the parameters from simulated data. Let SI be the set of indices for the serial intervals. In our example, we are estimating over one year, comprising 26 two-week serial intervals. The reported cases (known input) are given by C_i, and the variable ε_i^I is the residual between the measured and calculated cases. The full problem formulation is given by,

$$\min \sum_{i \in SI} \left(\varepsilon_i^I \right)^2$$

$$I_i = \frac{\beta I_{i-1}^\alpha S_{i-1}}{N} \ \forall \ i \in SI \setminus \{1\}$$

$$S_i = S_{i-1} - I_i \ \forall i \in SI \setminus \{1\}$$

$$C_i = I_i + \varepsilon_i^I$$

$$0 \le I_i, S_i \le N$$

$$0.5 \le \beta \le 70$$

$$0.5 \le \alpha \le 1.5$$

The following listing shows an abstract model for this nonlinear least-squares estimation problem:

```python
# disease_estimation.py
import pyomo.environ as pyo

model = pyo.AbstractModel()

model.S_SI = pyo.Set(ordered=True)

model.P_REP_CASES = pyo.Param(model.S_SI)
model.P_POP = pyo.Param()

model.I = pyo.Var(model.S_SI, bounds=(0,model.P_POP), \
    initialize=1)
model.S = pyo.Var(model.S_SI, bounds=(0,model.P_POP), \
    initialize=300)
model.beta = pyo.Var(bounds=(0.05, 70))
model.alpha = pyo.Var(bounds=(0.5, 1.5))
model.eps_I = pyo.Var(model.S_SI, initialize=0.0)

def _objective(model):
    return sum((model.eps_I[i])**2 for i in model.S_SI)
model.objective = pyo.Objective(rule=_objective, \
    sense=pyo.minimize)

def _InfDynamics(model, i):
    if i != 1:
        return model.I[i] == (model.beta * model.S[i-1] * \
            model.I[i-1]**model.alpha)/model.P_POP
    return pyo.Constraint.Skip

model.InfDynamics = pyo.Constraint(model.S_SI, rule=_InfDynamics)

def _SusDynamics(model, i):
    if i != 1:
        return model.S[i] == model.S[i-1] - model.I[i]
    return pyo.Constraint.Skip
model.SusDynamics = pyo.Constraint(model.S_SI, rule=_SusDynamics)

def _Data(model, i):
```

```
    return model.P_REP_CASES[i] == model.I[i]+model.eps_I[i]
model.Data = pyo.Constraint(model.S_SI, rule=_Data)

# create the ConcreteModel
instance = model.create_instance('disease_estimation.dat')
status = pyo.SolverFactory('ipopt').solve(instance)
pyo.assert_optimal_termination(status)

print(' ***')
print(' *** Optimal beta Value: %.2f' % pyo.value(instance.beta))
print(' *** Optimal alpha Value: %.2f' % \
    pyo.value(instance.alpha))
print(' ***')
```

The Pyomo data file containing the data for an instance of this model is given by:

```
# disease_estimation.dat

set S_SI := 1 2 3 4 5 6 7 8 9 10 11 12 13 14
            15 16 17 18 19 20 21 22 23 24 25 26 ;

param P_POP := 300.000000;

param P_REP_CASES default 0.0 :=
   1   1.000000
   2   2.000000
   3   4.000000
   4   8.000000
   5   15.000000
   6   27.000000
   7   44.000000
   8   58.000000
   9   55.000000
  10   32.000000
  11   12.000000
  12   3.000000
  13   1.000000
;
```

We can solve the estimation problem with the command:

```
python disease_estimation.py
```

and it will produce output similar to the following:

```
***
*** Optimal beta Value: 1.99
*** Optimal alpha Value: 1.00
***
```

We generated the data with $\beta=2$ and $\alpha=1$, so these results look quite reasonable.

7.4.4 Reactor Design

Chemical reactors are often the most important unit operations in a chemical plant. Reactors come in many forms, however two of the most common idealizations are the continuously stirred tank reactor (CSTR) and the plug flow reactor. The CSTR is often used in modeling studies, and it can be effectively modeled as a lumped parameter system. In this example, we will consider the following reaction scheme known as the Van de Vusse reaction:

$$A \xrightarrow{k_1} B \xrightarrow{k_2} C$$

$$2A \xrightarrow{k_3} D$$

A diagram of the system is shown in Figure 7.2, where F is the volumetric flowrate. The reactor is assumed to be filled to a constant volume, and the mixture is assumed to have constant density, so the volumetric flowrate into the reactor is equal to the volumetric flowrate out of the reactor. Since the reactor is assumed to be well-mixed, the concentrations in the reactor are equivalent to the concentrations of each component flowing out of the reactor, given by c_A, c_B, c_C, and c_D.

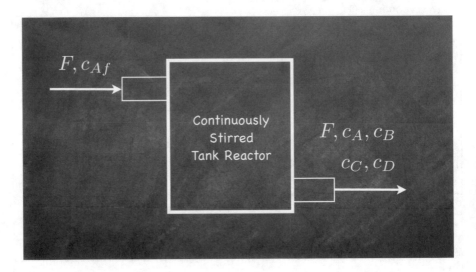

Fig. 7.2: Continuously stirred tank reactor system producing desired product B, and undesired products C and D from reactant A.

Consider the following reactor problem adapted from Bequette [7]. The goal is to produce product B from a feed containing reactant A. If we design a reactor that is too small, we will obtain insufficient conversion of A to the desired product B. However, given the reaction scheme, if the reactor is too large (e.g., too much reaction is allowed to occur), a significant amount of the desired product B will be further reacted to form the undesired product C. As a result, the goal in this exercise will be to solve for the optimal reactor volume producing the maximum outlet concentration for product B.

The steady-state mole balances for each of the four components are given by,

$$0 = \frac{F}{V}c_{Af} - \frac{F}{V}c_A - k_1 c_A - 2k_3 c_A^2$$

$$0 = -\frac{F}{V}c_B + k_1 c_A - k_2 c_B$$

$$0 = -\frac{F}{V}c_C + k_2 c_B$$

$$0 = -\frac{F}{V}c_D + k_3 c_A^2$$

The known parameters for the system are,

$$c_{Af} = 10000 \ \frac{\text{gmol}}{\text{m}^3} \quad k_1 = \frac{5}{6} \ \text{min}^{-1} \quad k_2 = \frac{5}{3} \ \text{min}^{-1} \quad k_3 = \frac{1}{6000} \ \frac{\text{m}^3}{\text{gmol min}}.$$

Since the volumetric flowrate F always appears as the numerator over the reactor volume V, it is common to consider this ratio as a single variable, called the space-velocity (sv). Our optimization formulation will seek to find the space-velocity that maximizes the outlet concentration of the desired product B.

The following file includes a function that builds the concrete model for the reactor design problem as well as code that will solve the problem if the Python file is executed directly:

```
import pyomo.environ
import pyomo.environ as pyo

def create_model(k1, k2, k3, caf):
    # create the concrete model
    model = pyo.ConcreteModel()

    # create the variables
    model.sv = pyo.Var(initialize=1.0, within=pyo.PositiveReals)
    model.ca = pyo.Var(initialize=5000.0, within=pyo.PositiveReals)
    model.cb = pyo.Var(initialize=2000.0, within=pyo.PositiveReals)
    model.cc = pyo.Var(initialize=2000.0, within=pyo.PositiveReals)
    model.cd = pyo.Var(initialize=1000.0, within=pyo.PositiveReals)

    # create the objective
    model.obj = pyo.Objective(expr = model.cb, sense=pyo.maximize)

    # create the constraints
```

```
    model.ca_bal = pyo.Constraint(expr = (0 == model.sv * caf \
                   - model.sv * model.ca - k1 * model.ca \
                   - 2.0 * k3 * model.ca ** 2.0))

    model.cb_bal = pyo.Constraint(expr=(0 == -model.sv * model.cb \
                   + k1 * model.ca - k2 * model.cb))

    model.cc_bal = pyo.Constraint(expr=(0 == -model.sv * model.cc \
                   + k2 * model.cb))

    model.cd_bal = pyo.Constraint(expr=(0 == -model.sv * model.cd \
                   + k3 * model.ca ** 2.0))

    return model

if __name__ =='__main__':
    # solve a single instance of the problem
    k1 = 5.0/6.0 # min^-1
    k2 = 5.0/3.0 # min^-1
    k3 = 1.0/6000.0 # m^3/(gmol min)
    caf = 10000.0 # gmol/m^3

    m = create_model(k1, k2, k3, caf)
    status = pyo.SolverFactory('ipopt').solve(m)
    pyo.assert_optimal_termination(status)
    m.pprint()
```

The problem can be solved with command:

```
python ReactorDesign.py
```

The optimal space-velocity is 1.34, giving an outlet concentration for B of 1072.

It is easy to construct the model and execute a script using the model. For example, if we wanted to solve this design problem for different values of the feed concentration, we could use the following code:

```
import pyomo.environ as pyo
from ReactorDesign import create_model

# set the data (native Python data)
k1 = 5.0/6.0 # min^-1
k2 = 5.0/3.0 # min^-1
k3 = 1.0/6000.0 # m^3/(gmol min)

# solve the model for different values of caf and report results
print('{:>10s}\t{:>10s}\t{:>10s}'.format('CAf', 'SV', 'CB'))
for cafi in range(1,11):
    caf = cafi*1000.0 # gmol/m^3

    # create the model with the new data
    # note, we could do this more efficiently with
    # mutable parameters
    m = create_model(k1, k2, k3, caf)

    # solve the problem
```

```
status = pyo.SolverFactory('ipopt').solve(m)
print("{:10g}\t{:10g}\t{:10g}".\
    format(caf, pyo.value(m.sv), pyo.value(m.cb)))
```

This script can be executed using the `python` command, producing the following results:

CAf	SV	CB
1000	1.21294	157.564
2000	1.23903	294.346
3000	1.25993	416.943
4000	1.27729	529.051
5000	1.29209	632.993
6000	1.30495	730.339
7000	1.31629	822.212
8000	1.32641	909.447
9000	1.33553	992.687
10000	1.34381	1072.44

This example illustrates the scripting capabilities of Pyomo. See Chapter 5 for additional scripting examples and further description of these capabilities.

Chapter 8
Structured Modeling with Blocks

Abstract This chapter documents how to express hierarchically-structured models using Pyomo's `Block` component. Many models contain significant hierarchical structure; that is, they are composed of repeated groups of conceptually related modeling components. Pyomo allows the modeler to define fundamental building blocks, and then construct the overall problem by connecting these building blocks together in an object-oriented manner. In this chapter, we describe the fundamental `Block` component along with common examples of its use, including repeated components and managing model scope.

8.1 Introduction

Optimization solvers typically rely on getting a model in a standardized form. For example, linear solvers accept models built on a standard form similar to:

$$\min c^T x$$
$$\text{s.t.} \quad Ax \le b \ .$$
$$x \ge 0$$

Here, the variables are lumped together into a single vector x, and constraints are represented in simplified matrix form. While this form is convenient for algorithms directly manipulating these matrices, it is not an easy form for a modeler to generate, manipulate, or debug. Algebraic Modeling Languages (AMLs) directly address this challenge by allowing modelers to provide distinguishing names to modeling components (e.g., variables or constraints) and to define the model over index sets. Since models are often composed of repeated mathematical expressions, this allows the expression of large models with relatively few lines of code, which are also easier to document, understand, modify, and debug.

As models become larger and more complex, we often want to carry this concept further using the principle of *composition* from objet-oriented programming.

© The Author(s), under exclusive license to Springer Nature Switzerland AG 2021
M. L. Bynum et al., *Pyomo — Optimization Modeling in Python*, Springer Optimization
and Its Applications 67, https://doi.org/10.1007/978-3-030-68928-5_8

In this approach, we group - or *compose* - variables and constraints that are conceptually related into a single object. This modeling approach, does not emphasize that the constraints are connected by a common expression generator, but rather the variables and constraints describe a certain (often physical) concept. For example, the group of variables and constraints could represent the operating behavior of an electric generator (ramp-up limits, ramp-down limits, cost curves) or chemical process equipment like a distillation column (the mass, equilibrium, and energy balance equations). Other examples include multi-period optimization problems where the same fundamental model is repeated over many time periods, or stochastic programming problems where the same basic model is repeated over different scenarios with different parameters. In Pyomo, we use the `Block` component to support general composition of modeling components like those described.

The `Block` component is a container for organizing groups of variables and constraints, and it can contain any number of named Pyomo components in exactly the same way models do. In fact, `ConcreteModel` and `AbstractModel` are themselves special implementations of the `Block` component. `Block` components can be added to a model or another block, allowing modelers to construct hierarchical model structures based on fundamental building blocks in an object-oriented manner.

This concept is illustrated in Figure 8.1, which shows one possible block-oriented representation of a an electrical grid model. In this example, individual blocks define

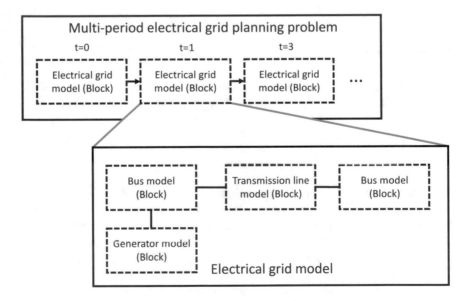

Fig. 8.1: The electrical grid model can be composed from individual blocks representing the generators, buses, and transmission lines. Furthermore, a multi-period model can be constructed in a hierarchical manner treating the electrical grid model as its fundamental building blocks.

the necessary variables and equations describing a single piece of equipment, be it a generator, bus, or transmission line. Then, blocks are connected together to form the entire (single time period) electrical grid power flow model. This model could be solved on its own. However, it can also be used as a building block for a higher-level model, like the multi-period planning problem illustrated in this figure. In this manner, Pyomo can represent very complex models built using smaller, less complex, reusable pieces.

Blocks provide both modelers and alorithm developers several useful benefits:

composition Modelers can build - and test - small models or model components and then assemble them into larger, more complex system models.

annotation The model can explicitly annotate model structure and provide "hints" to both algorithms (e.g., decomposition algorithms like Benders decomposition or progressive hedging) and transformations (e.g., generalized disjunctive programming) for how to manipulate the model.

sandboxing Because each Block is its own container, it enables both model (component) developers and algorithm developers to build and manipulate modeling components in their own "private" sandboxes, without worrying about naming collisions or other interference from other developers.

8.2 Block structures

The Pyomo `Block` component can be treated in much the same way as a model: components are added directly to the block as attributes. Since `Block` components may contain any other Pyomo modeling components, including other blocks, it is possible to construct arbitrarily nested hierarchical structures.

The following code snippet shows a basic block hierarchy:

```
model = pyo.ConcreteModel()
model.x = pyo.Var()
model.P = pyo.Param(initialize=5)
model.S = pyo.RangeSet(model.P)
model.b = pyo.Block()
model.b.I = pyo.RangeSet(model.P)
model.b.x = pyo.Var(model.b.I)
model.b.y = pyo.Var(model.S)
model.b.b = pyo.Block([1,2])
model.b.b[1].x = pyo.Var()
model.b.b[2].x = pyo.Var()
```

Here, `model` contains a variable (`model.x`), a parameter (`model.P`), and a set (`model.S`). It also contains a `Block` given by `model.b`. This block itself contains a set, two variables, and another indexed block. Notice components in a block can reference components in other blocks or in the parent model. For example, the variable `model.b.x` uses a set from its parent block, and the variable `model.b.y` references a set from the parent of the parent block (in this case, the model). Components and expressions can reference components from anywhere within the hierarchy.

NOTE: Within one block, Pyomo supports references to components from any other block. However, it is generally good object-oriented practice to only reference components from the current or lower levels in the hierarchy. This promotes resusability of your blocks within other models without strong assumptions about the structure of the owning or parent blocks.

Note in the code snippet each block defines its own *component namespace*; while component names must be unique within a single block, they do not need to be globally unique. This allows blocks to be constructed safely without concerns about definitions in one block colliding or interfering with definitions in other blocks. This leads to all components having two forms of a name: the local name, which may be repeated elsewhere in the model, and a globally-unique *fully qualified* name that includes the names of the parent block(s) separated by periods:

```
print(model.x.local_name)          # x
print(model.x.name)                # x
print(model.b.x.local_name)        # x
print(model.b.x.name)              # b.x
print(model.b.b[1].x.local_name)   # x
print(model.b.b[1].x.name)         # b.b[1].x
```

When discussing the block hierarchy, we adopt terminology from tree structures and refer to the block one level up the hierarchy (toward the top-level model) as the *parent block*, and all components contained within a block as its *children*. The *root* of the block hierarchy is always the current model. Pyomo components all provide a set of standard methods for moving around the component hierarchy:

parent_component() Each modeling object in a Pyomo model is owned by a single component. Calling parent_component() on a modeling object returns the *component* that holding (or owning) the modeling object. For members of an indexed component, parent_component() returns the indexed component object containing the component. For scalar components, the parent component is the scalar component itself.

parent_block() Each modeling component is attached to a single block. Calling parent_block() on a Pyomo modeling object returns the block that owns the object's parent_component().

model() Calling model() on a modeling object walks up the parent_block() calls to return the top-level (root) block.

getitem As with models, child components can be accessed programmatically by attribute lookup on the block.

component() Child components can also be retrieved by name using the Blocks component() method.

The following illustrates how to move around the component hierarchy:

```
model.b.b[1].x.parent_component()    # is model.b.b[1].x
model.b.b[1].x.parent_block()        # is model.b.b[1]
model.b.b[1].x.model()               # is model
model.b.b[1].component('x')          # is model.b.b[1].x
model.b.x[1].parent_component()      # is model.b.x
model.b.x[1].parent_block()          # is model.b
model.b.x[1].model()                 # is model
model.b.component('x')               # is model.b.x
```

Block components can also be created and populated, and then later added to a model. The code below shows creation of a block that is later added to a model. It also illustrates how a parent model can make use of sets and parameters contained in a child block.

```
new_b = pyo.Block()
new_b.x = pyo.Var()
new_b.P = pyo.Param(initialize=5)
new_b.I = pyo.RangeSet(10)

model = pyo.ConcreteModel()
model.b = new_b
model.x = pyo.Var(model.b.I)
```

NOTE: In this example, the `Block` object `new_b` is not initialized when it is declared. In this manner, it is abstract until it is added to the `ConcreteModel` object. At this point, it is immediately initialized. Similarly, when a `Block` object is added to an `AbstractModel`, it is not initialized until the owning abstract model object is initialized.

8.3 Blocks as Indexed Components

As with other Pyomo components, `Block` components may also be indexed and initialized using a construction rule. However, block construction rules follow a slightly different convention: the first argument to a block rule is the block to be populated rather than the owning block. This block has already been attached to the model so methods like `model()` and `parent_block()` will work as expected. Within a rule, one can either directly populate the block by assigning components to it, or create a new block and return it from the rule.

The following example illustrates the use of construction rules for blocks:

```
model = pyo.ConcreteModel()
model.P = pyo.Param(initialize=3)
model.T = pyo.RangeSet(model.P)

def xyb_rule(b, t):
    b.x = pyo.Var()
    b.I = pyo.RangeSet(t)
    b.y = pyo.Var(b.I)
    b.c = pyo.Constraint(expr=b.x == 1 - sum(b.y[i] for i in b.I))
model.xyb = pyo.Block(model.T, rule=xyb_rule)
```

Here, an indexed `Block` component containing one block for each element of the set `model.T` is defined. In the construction rule, we create two variables and a set. These construction rules are just as flexible as those for other components. The set `b.I` created in this example is different for each block, and consequently the variable `b.y` is also a different length for each block. This illustrates another feature of blocks. Often, different blocks contain exactly the same model structure, just with different data. It is also possible to construct blocks with a different structure based on data available in the construction rule.

This example can be extended to print the constraint body `c` for each of the blocks:

```
for t in model.T:
    print(model.xyb[t].c.body)
```

The constraints expand appropriately and they contain fully qualified names of variables in each of the subblocks:

```
-1.0 + xyb[1].x + xyb[1].y[1]
-1.0 + xyb[2].x + xyb[2].y[1] + xyb[2].y[2]
-1.0 + xyb[3].x + xyb[3].y[1] + xyb[3].y[2] + xyb[3].y[3]
```

8.4 Construction Rules within Blocks

Like model objects, blocks can contain other modeling components, including `Set` and `Param` objects. Additionally, blocks can be initialized with modeling components that are themselves constructed using rules. However, doing this exposes a subtelty of Pyomo component construction rules.

Up to this point we have frequently referred to the first argument of a component rule as the "model", but this is not completely correct. The first argument to component rules is actually the *owning block* of the component being constructed. For "flat" models (models without any sub-blocks), the owning block is indeed the model, but this will not be the case for hierarchically-structured models. If needed, the model object can be obtained from the owning block using the aforementioned `model()` method.

For example, consider the following alternative declaration of the xyb block from the previous example:

```
def xyb_rule(b, t):
    b.x = pyo.Var()
    b.I = pyo.RangeSet(t)
    b.y = pyo.Var(b.I, initialize=1.0)
    def _b_c_rule(_b):
        return _b.x == 1.0 - sum(_b.y[i] for i in _b.I)
    b.c = pyo.Constraint(rule=_b_c_rule)
model.xyb = pyo.Block(model.T, rule=xyb_rule)
```

In this example, the xyb block includes a constraint that is defined with the rule _b_c_rule. The owning block _b passed to this rule is the same as b. However, the _b variable is defined locally. This allows the rule to be used even if the owning block is constructed in a different manner.

Since the owning block (or model) is NOT passed into a block construction rule, the modeler may need another mechanism to access components on the parent or other blocks in the hierarchy. The component methods parent_block and model facilitate moving up the block hierarchy. The parent_block() of any component or component data object is the block that the component is attached to. The model() method on any component or component data object returns the block object at the root of the tree.

8.5 Extracting values from hierarchical models

While blocks support a convenient mechanism for expressing composite concepts (e.g., a time-period, a scenario), this results in some data becoming more spread out across your model. However, we can access the components by explicitly iterating over blocks and their associated variables:

```
for t in model.xyb:
  for i in model.xyb[t].y:
    print("%s %f" % (model.xyb[t].y[i], \
        pyo.value(model.xyb[t].y[i])))
```

Additionally, Pyomo's slice notation can be used to dynamically extract a subset of the blocks or variable values:

```
for y in model.xyb[:].y[:]:
    print("%s %f" % (y, pyo.value(y)))
```

8.6 Blocks Example: Optimal Multi-Period Lot-Sizing

We now demonstrate a complete model based on blocks using a well-known multi-period optimization problem for optimal lot-sizing [31]. Our goal in the lot-sizing problem is to determine the optimal production X_t in each time period $t \in T$ given known demands d_t. We let y_t be a binary variable indicating whether or not there is any production in time period t, and assume that there is a fixed cost c_t if we decide to produce in time period t. The inventory I_t at the end of each time period is a function of the previous inventory, production, and sales,

$$I_t = I_{t-1} + X_t - d_t.$$

If we allow the inventory to be negative (meaning we did not meet demands and we have a backlog of orders), we can represent the inventory as $I_t = I_t^+ - I_t^-$ where we restrict both I_t^+ and I_t^- to be non-negative. Here, I_t^+ represents inventory that we are holding, and I_t^- represents a backlog of orders. We can assign an inventory holding cost and a shortage cost (cost of keeping a backlog) as h_t^+ and h_t^-, respectively.

With this description, the optimization problem can be formulated as,

$$\min \sum_{t \in T} c_t y_t + h_t^+ I_t^+ + h_t^- I_t^- \tag{LS.1}$$

$$\text{s.t.} \quad I_t = I_{t-1} + X_t - d_t \qquad \forall t \in T \qquad \text{(LS.2)}$$

$$I_t = I_t^+ - I_t^- \qquad \forall t \in T \qquad \text{(LS.3)}$$

$$X_t \leq P y_t \qquad \forall t \in T \qquad \text{(LS.4)}$$

$$X_t, I_t^+, I_t^- \geq 0 \qquad \forall t \in T \qquad \text{(LS.5)}$$

$$y_t \in \{0,1\} \qquad \forall t \in T \qquad \text{(LS.6)}$$

where Equation (LS.4) is a constraint that only allows production in time period t if the indicator variable $y_t = 1$. The data for our problem is provided in Table 8.1.

Table 8.1: Data for Lot-Sizing Problem

Parameter	Description	Value
c	fixed cost of production	4.6
I_0^+	initial value of positive inventory	5.0
I_0^-	initial value of backlogged orders	0.0
h^+	cost (per unit) of holding inventory	0.7
h^-	shortage cost (per unit)	1.2
P	maximum production amount (big-M value)	5
d	demand	[5, 7, 6.2, 3.1, 1.7]

8.6.1 A Formulation Without Blocks

We can formulate the lot-sizing problem *without* blocks using the Pyomo code
shown below.

```python
import pyomo.environ as pyo

model = pyo.ConcreteModel()
model.T = pyo.RangeSet(5) # time periods

i0 = 5.0 # initial inventory
c = 4.6 # setup cost
h_pos = 0.7 # inventory holding cost
h_neg = 1.2 # shortage cost
P = 5.0 # maximum production amount

# demand during period t
d = {1: 5.0, 2:7.0, 3:6.2, 4:3.1, 5:1.7}

# define the variables
model.y = pyo.Var(model.T, domain=pyo.Binary)
model.x = pyo.Var(model.T, domain=pyo.NonNegativeReals)
model.i = pyo.Var(model.T)
model.i_pos = pyo.Var(model.T, domain=pyo.NonNegativeReals)
model.i_neg = pyo.Var(model.T, domain=pyo.NonNegativeReals)

# define the inventory relationships
def inventory_rule(m, t):
    if t == m.T.first():
        return m.i[t] == i0 + m.x[t] - d[t]
    return m.i[t] == m.i[t-1] + m.x[t] - d[t]
model.inventory = pyo.Constraint(model.T, rule=inventory_rule)

def pos_neg_rule(m, t):
    return m.i[t] == m.i_pos[t] - m.i_neg[t]
model.pos_neg = pyo.Constraint(model.T, rule=pos_neg_rule)

# create the big-M constraint for the production indicator variable
def prod_indicator_rule(m,t):
    return m.x[t] <= P*m.y[t]
model.prod_indicator = pyo.Constraint(model.T, rule=prod_indicator_rule)

# define the cost function
def obj_rule(m):
    return sum(c*m.y[t] + h_pos*m.i_pos[t] + h_neg*m.i_neg[t] for t in m.T)
model.obj = pyo.Objective(rule=obj_rule)

# solve the problem
solver = pyo.SolverFactory('glpk')
solver.solve(model)

# print the results
for t in model.T:
    print('Period: {0}, Prod. Amount: {1}'.format(t, pyo.value(model.x[t])))
```

This example uses standard Pyomo syntax as discussed in early chapters of the
book. If we were considering the lot-sizing problem over a single time period only,
our variable declarations would have looked like,

```
# define the variables
model.y = pyo.Var(domain=pyo.Binary)
model.x = pyo.Var(domain=pyo.NonNegativeReals)
model.i = pyo.Var()
model.i_pos = pyo.Var(domain=pyo.NonNegativeReals)
model.i_neg = pyo.Var(domain=pyo.NonNegativeReals)
```

In the multi-period case, we have the same fundamental variables and constraints defined over each time period. Here, the variable declarations looked like,

```
# define the variables
model.y = pyo.Var(model.T, domain=pyo.Binary)
model.x = pyo.Var(model.T, domain=pyo.NonNegativeReals)
model.i = pyo.Var(model.T)
model.i_pos = pyo.Var(model.T, domain=pyo.NonNegativeReals)
model.i_neg = pyo.Var(model.T, domain=pyo.NonNegativeReals)
```

If we were considering a multi-period and multi-scenario problem (e.g., a stochastic programming formulation for lot-sizing under uncertainty), the variable declarations would have looked like,

```
# define the variables
model.y = pyo.Var(model.T, model.S, domain=pyo.Binary)
model.x = pyo.Var(model.T, model.S, domain=pyo.NonNegativeReals)
model.i = pyo.Var(model.T, model.S,)
model.i_pos = pyo.Var(model.T, model.S, \
    domain=pyo.NonNegativeReals)
model.i_neg = pyo.Var(model.T, model.S, \
    domain=pyo.NonNegativeReals)
```

In each of these examples, when we add new complexity, or an additional *layer* onto the model, we add a new index to the variables and constraints. This approach is very common in the field of operations research. Unfortunately, it requires completely redefining the model with each new layer, and it does not readily support construction of hierarchical models with reusable code. Blocks provide another approach to easily support model reuse in an object-oriented fashion.

8.6.2 A Formulation With Blocks

We now show how blocks can be used to write this problem. Most of the constraints in the multi-period lot-sizing problem are defined over $t \in T$, and they can be logically grouped together by time. Pyomo allows us to define blocks, each with the variables and constraints for a single time period only, and then link them together to form the overall model.

Considering the lot-sizing problem again, the variables and constraints within a rule can be defined to provide a block for a single period of the lot-sizing problem:

```
# create a block for a single time period
def lotsizing_block_rule(b, t):
    # define the variables
    b.y = pyo.Var(domain=pyo.Binary)
    b.x = pyo.Var(domain=pyo.NonNegativeReals)
    b.i = pyo.Var()
    b.i0 = pyo.Var()
    b.i_pos = pyo.Var(domain=pyo.NonNegativeReals)
    b.i_neg = pyo.Var(domain=pyo.NonNegativeReals)

    # define the constraints
    b.inventory = pyo.Constraint(expr=b.i == b.i0 + b.x - d[t])
    b.pos_neg = pyo.Constraint(expr=b.i == b.i_pos - b.i_neg)
    b.prod_indicator = pyo.Constraint(expr=b.x <= P * b.y)
model.lsb = pyo.Block(model.T, rule=lotsizing_block_rule)
```

Here, the variables and constraints for a single time period t are defined within the rule. The `Block` component is then indexed over the set `model.T`, and the declaration constructs a lot-sizing block for each entry in `model.T`. The `model` object now contains a block for each time period t. The final action is to provide constraints linking the blocks together (setting the initial inventory of one block equal to the final inventory of the previous block), and to define the objective function over all the blocks. The full code listing for the block version of the multi-period lot-sizing problem is shown below.

```
import pyomo.environ as pyo

model = pyo.ConcreteModel()
model.T = pyo.RangeSet(5) # time periods

i0 = 5.0 # initial inventory
c = 4.6 # setup cost
h_pos = 0.7 # inventory holding cost
h_neg = 1.2 # shortage cost
P = 5.0 # maximum production amount

# demand during period t
d = {1: 5.0, 2:7.0, 3:6.2, 4:3.1, 5:1.7}

# create a block for a single time period
def lotsizing_block_rule(b, t):
    # define the variables
    b.y = pyo.Var(domain=pyo.Binary)
    b.x = pyo.Var(domain=pyo.NonNegativeReals)
    b.i = pyo.Var()
    b.i0 = pyo.Var()
    b.i_pos = pyo.Var(domain=pyo.NonNegativeReals)
    b.i_neg = pyo.Var(domain=pyo.NonNegativeReals)

    # define the constraints
    b.inventory = pyo.Constraint(expr=b.i == b.i0 + b.x - d[t])
    b.pos_neg = pyo.Constraint(expr=b.i == b.i_pos - b.i_neg)
    b.prod_indicator = pyo.Constraint(expr=b.x <= P * b.y)
```

```python
model.lsb = pyo.Block(model.T, rule=lotsizing_block_rule)

# link the inventory variables between blocks
def i_linking_rule(m, t):
    if t == m.T.first():
        return m.lsb[t].i0 == i0
    return m.lsb[t].i0 == m.lsb[t-1].i
model.i_linking = pyo.Constraint(model.T, rule=i_linking_rule)

# construct the objective function over all the blocks
def obj_rule(m):
    return sum(c*m.lsb[t].y + h_pos*m.lsb[t].i_pos + \
        h_neg*m.lsb[t].i_neg for t in m.T)
model.obj = pyo.Objective(rule=obj_rule)

### solve the problem
solver = pyo.SolverFactory('glpk')
solver.solve(model)

# print the results
for t in model.T:
    print('Period: {0}, Prod. Amount: {1}'.format(t, \
        pyo.value(model.lsb[t].x)))
```

This formulation is small, so it can be difficult to see the benefit of blocks. However, as models grow in size and complexity, this object-oriented modeling concept allows us to define small pieces of the model in self-contained chunks of code, and then build the large model by pulling these pieces together. This example was selected in part because it is a heavily studied, classic multi-stage inventory model. One can easily imagine extensions to the model to include additional constraints and costs. In fact, many such models have appeared in the academic literature and in practical application. In large models, it is common to write methods or classes to define individual blocks and reuse code within several different, high-level optimization formulations.

Chapter 9
Performance: Model Construction and Solver Interfaces

Abstract This chapter documents tools for profiling model construction and improving the performance of both model construction and interaction with solvers. We begin by discussing various profiling tools which can be used to help identify performance bottlenecks. Pyomo has built-in profiling capabilities, but there are also Python packages, such as cProfile and line_profiler, dedicated to performance profiling. Section 9.2 discusses the LinearExpression class, which can be used to drastically improve model construction time for some applications. Section 9.3 describes how persistent solver interfaces can be used to repeatedly solve models with small changes very efficiently. Finally, Section 9.4 discusses sparse index sets.

9.1 Profiling to Identify Performance Bottlenecks

When performance is an issue, it is necessary to identify the performance bottlenecks before improvements can be made. This section describes various profiling tools which can be used to identify performance bottlenecks.

Throughout this section, we will be revisiting the warehouse location problem as an example. The warehouse location problem was first introduced in Section 3.2. Here, we will work with a continuous relaxation of the problem for illustration. We rewrite the warehouse location problem so the model is built in a function and the maximum number of warehouses is a mutable parameter.

© The Author(s), under exclusive license to Springer Nature Switzerland AG 2021
M. L. Bynum et al., *Pyomo — Optimization Modeling in Python*, Springer Optimization
and Its Applications 67, https://doi.org/10.1007/978-3-030-68928-5_9

```python
import pyomo.environ as pyo # import pyomo environment
import cProfile
import pstats
import io
from pyomo.common.timing import TicTocTimer, report_timing
from pyomo.opt.results import assert_optimal_termination
from pyomo.core.expr.numeric_expr import LinearExpression
import matplotlib.pyplot as plt
import numpy as np
np.random.seed(0)
```

```python
def create_warehouse_model(num_locations=50, num_customers=50):
    N = list(range(num_locations)) # warehouse locations
    M = list(range(num_customers)) # customers

    d = dict() # distances from warehouse locations to customers
    for n in N:
        for m in M:
            d[n, m] = np.random.randint(low=1, high=100)
    max_num_warehouses = 2

    model = pyo.ConcreteModel(name="(WL)")
    model.P = pyo.Param(initialize=max_num_warehouses,
                        mutable=True)

    model.x = pyo.Var(N, M, bounds=(0, 1))
    model.y = pyo.Var(N, bounds=(0, 1))

    def obj_rule(mdl):
        return sum(d[n,m]*mdl.x[n,m] for n in N for m in M)
    model.obj = pyo.Objective(rule=obj_rule)

    def demand_rule(mdl, m):
        return sum(mdl.x[n,m] for n in N) == 1
    model.demand = pyo.Constraint(M, rule=demand_rule)

    def warehouse_active_rule(mdl, n, m):
        return mdl.x[n,m] <= mdl.y[n]
    model.warehouse_active = pyo.Constraint(N, M, \
        rule=warehouse_active_rule)

    def num_warehouses_rule(mdl):
        return sum(mdl.y[n] for n in N) <= model.P
    model.num_warehouses = \
        pyo.Constraint(rule=num_warehouses_rule)

    return model
```

9.1.1 Report Timing

Pyomo has a very useful function, `report_timing`, for profiling model construction. The following illustrates how to use `report_timing` to profile the construction of the warehouse location problem.

```
report_timing()
print('Building model')
print('--------------')
m = create_warehouse_model(num_locations=200, num_customers=200)
```

The output of the code is shown next.

```
Building model
--------------
        0 seconds to construct Block ConcreteModel; 1 index total
        0 seconds to construct Set Any; 1 index total
        0 seconds to construct Param P; 1 index total
        0 seconds to construct Set OrderedSimpleSet; 1 index total
        0 seconds to construct Set OrderedSimpleSet; 1 index total
        0 seconds to construct Set SetProduct_OrderedSet; 1 index total
        0 seconds to construct Set SetProduct_OrderedSet; 1 index total
     0.09 seconds to construct Var x; 40000 indicies total
        0 seconds to construct Set OrderedSimpleSet; 1 index total
        0 seconds to construct Var y; 200 indicies total
     0.19 seconds to construct Objective obj; 1 index total
        0 seconds to construct Set OrderedSimpleSet; 1 index total
     0.11 seconds to construct Constraint demand; 200 indicies total
        0 seconds to construct Set OrderedSimpleSet; 1 index total
        0 seconds to construct Set OrderedSimpleSet; 1 index total
        0 seconds to construct Set SetProduct_OrderedSet; 1 index total
        0 seconds to construct Set SetProduct_OrderedSet; 1 index total
     0.62 seconds to construct Constraint warehouse_active; 40000 indicies total
        0 seconds to construct Constraint num_warehouses; 1 index total
```

Calling the `report_timing` function causes Pyomo to print the time required to build each component. In the second to last line in the output, we can see building the `warehouse_active` constraint accounts for the majority of the model construction time. Note that any data processing done inside the constraint rules is included, so not all of the time is necessarily spent in Pyomo. In this example, there is not any data processesing within the rule, so we know building the expressions for the `warehouse_active` constraint is the bottleneck.

9.1.2 TicTocTimer

Pyomo also provides a `TicTocTimer` for convenient timing. In the following example, we compare the time required to build the model and the time to write an LP file and solve the problem with Gurobi. Suppose we have this function to solve the model:

```
def solve_warehouse_location(m):
    opt = pyo.SolverFactory('gurobi')
    res = opt.solve(m)
    assert_optimal_termination(res)
```

We could use the following script to use the `TicTocTimer` for timing:

```
timer = TicTocTimer()
timer.tic('start')
```

```
m = create_warehouse_model(num_locations=200, num_customers=200)
timer.toc('Built model')
solve_warehouse_location(m)
timer.toc('Wrote LP file and solved')
```

The output of the code is shown next.

```
[      0.00] start
[+    1.22] Built model
[+    4.05] Wrote LP file and solved
```

The output states it took 1.22 seconds to build the Pyomo model and a total of 4.05 seconds to write the LP file and solve the problem. By utilizing the `TicTocTimer`, we found that writing the LP file and solving the problem took significantly longer than constructing the model. However, it is not yet clear what fraction of the 4.05 seconds is spent writing the LP file, which is where `cProfile` is useful.

9.1.3 Profilers

Many Python packages dedicated to performance profiling exist. Two examples of such packages are `cProfile` (https://docs.python.org/3/library/profile.html) and `line_profiler` (https://github.com/pyutils/line_profiler). The `cProfile` package reports the time spent in each function and the `line_profiler` package reports the time spent on each line of a function.

Let's revisit the warehouse location problem as an example. First, we write a function to perform a parametric sweep, repeatedly solving the warehouse location problem while varying the maximum number of warehouses, `m.P`.

```
def solve_parametric():
    m = create_warehouse_model(num_locations=50, num_customers=50)
    opt = pyo.SolverFactory('gurobi')
    p_values = list(range(1, 31))
    obj_values = list()
    for p in p_values:
        m.P.value = p
        res = opt.solve(m)
        assert_optimal_termination(res)
        obj_values.append(res.problem.lower_bound)
```

We first call the `solve_parametric` function and report the time required to perform the parameter sweep.

```
solve_parametric()
timer.toc('Finished parameter sweep')
```

The output of the code is shown next.

```
[+ 7.28] finished parameter sweep
```

The `TicTocTimer` reports that it took 7.28 seconds to complete the parameter sweep. Next, we write a function to help print the output from `cProfile`.

```python
def print_c_profiler(pr, lines_to_print=15):
    s = io.StringIO()
    stats = pstats.Stats(pr, stream=s).sort_stats('cumulative')
    stats.print_stats(lines_to_print)
    print(s.getvalue())
    s = io.StringIO()
    stats = pstats.Stats(pr, stream=s).sort_stats('tottime')
    stats.print_stats(lines_to_print)
    print(s.getvalue())
```

We use the `pstats` package to sort the output from `cProfile` and print the specified number of lines. We print the statistics sorted by both cumulative time and total time. As described in the cProfile documentation, the cumulative time is the time spent in the corresponding function, including all functions called within the function. The total time is the time spent in the corresponding function, excluding all calls to functions within the specified function.

We can use the `cProfile` package as follows.

```python
pr = cProfile.Profile()
pr.enable()
solve_parametric()
pr.disable()
print_c_profiler(pr)
```

The terminal output showing timing that results from running this code is shown next.

```
    14897207 function calls (14894602 primitive calls) in 10.229 seconds

Ordered by: cumulative time
List reduced from 559 to 15 due to restriction <15>

ncalls tottime percall cumtime percall filename:lineno(function)
     1 0.001 0.001 10.229 10.229 wl.py:112(solve_parametric)
    30 0.006 0.000 10.107 0.337 /.../pyomo/pyomo/opt/base/solvers.py:513(solve)
    30 0.000 0.000 5.758 0.192 /.../pyomo/pyomo/solvers/plugins/solvers/GUROBI.py:189(_presolve)
    30 0.000 0.000 5.758 0.192 /.../pyomo/pyomo/opt/solver/shellcmd.py:189(_presolve)
    30 0.001 0.000 5.726 0.191 /.../pyomo/pyomo/opt/base/solvers.py:653(_presolve)
    30 0.000 0.000 5.725 0.191 /.../pyomo/pyomo/opt/base/solvers.py:721(_convert_problem)
    30 0.002 0.000 5.724 0.191 /.../pyomo/pyomo/opt/base/convert.py:31(convert_problem)
    30 0.001 0.000 5.692 0.190 /.../pyomo/pyomo/solvers/plugins/converter/model.py:43(apply)
    30 0.001 0.000 5.675 0.189 /.../pyomo/pyomo/core/base/block.py:1736(write)
    30 0.016 0.001 5.673 0.189 /.../pyomo/pyomo/repn/plugins/cpxlp.py:84(__call__)
    30 0.804 0.027 5.650 0.188 /.../pyomo/pyomo/repn/plugins/cpxlp.py:380(_print_model_LP)
    30 0.003 0.000 3.528 0.118 /.../pyomo/pyomo/opt/solver/shellcmd.py:224(_apply_solver)
    30 0.002 0.000 3.524 0.117 /.../pyomo/pyomo/opt/solver/shellcmd.py:290(_execute_command)
    30 0.008 0.000 3.522 0.117 /.../pyutilib/pyutilib/subprocess/processmngr.py:433(run_command)
    30 0.001 0.000 3.292 0.110 /.../pyutilib/pyutilib/subprocess/processmngr.py:829(wait)


    14897207 function calls (14894602 primitive calls) in 10.229 seconds

Ordered by: internal time
List reduced from 559 to 15 due to restriction <15>

ncalls tottime percall cumtime percall filename:lineno(function)
    30 3.285 0.110 3.285 0.110 {built-in method posix.waitpid}
    30 0.804 0.027 5.650 0.188 /.../pyomo/pyomo/repn/plugins/cpxlp.py:380(_print_model_LP)
 76560 0.423 0.000 0.534 0.000 /.../pyomo/pyomo/repn/standard_repn.py:433(_collect_sum)
 76560 0.419 0.000 0.700 0.000 /.../pyomo/pyomo/repn/plugins/cpxlp.py:181(_print_expr_canonical)
 76560 0.339 0.000 1.049 0.000 /.../pyomo/pyomo/repn/standard_repn.py:982(_generate_standard_repn)
306000 0.338 0.000 0.586 0.000 /.../pyomo/pyomo/core/base/set.py:581(bounds)
    30 0.252 0.008 0.375 0.013 /.../pyomo/pyomo/solvers/plugins/solvers/GUROBI.py:363(process_soln_file)
 76560 0.197 0.000 1.411 0.000 /.../pyomo/pyomo/repn/standard_repn.py:254(generate_standard_repn)
 76560 0.159 0.000 1.812 0.000 /.../pyomo/pyomo/repn/plugins/cpxlp.py:572(constraint_generator)
225090 0.157 0.000 0.206 0.000 /.../pyomo/pyomo/core/base/constraint.py:206(has_ub)
153060 0.148 0.000 0.256 0.000 /.../pyomo/pyomo/core/expr/symbol_map.py:82(createSymbol)
```

```
 77220 0.124 0.000 0.272 0.000 {built-in method builtins.sorted}
153000 0.123 0.000 0.454 0.000 /.../pyomo/pyomo/core/base/var.py:407(ub)
153000 0.122 0.000 0.457 0.000 /.../pyomo/pyomo/core/base/var.py:394(lb)
229530 0.116 0.000 0.222 0.000 /.../pyomo/pyomo/repn/plugins/cpxlp.py:41(_get_bound)
```

The first line of output shows the total number of function calls and the total time required to run the profiled code. Note the 10.229 seconds reported by cProfile is significantly longer than the 7.28 seconds reported by the TicTocTimer. Using cProfile does add some overhead to the code being profiled. Therefore, cProfile should be used to identify bottlenecks (rather than comparing two algorithms, for example). The first block of output is sorted by cumulative time. Each row shows the cumulative time spent in the function on the far right of the row. As expected, the entire 10.229 seconds is spent within the solve_parametric function. Of the total 10.229 seconds, 10.107 seconds are spent within the Pyomo call to solve. We can already conclude very little time is spent constructing the model, changing the value of m.P, and checking the termination condition of the solver. Additionally, we see the call to write (which is where the LP file is written) takes 5.675 seconds of the 10.107 seconds spent in solve. Only 3.528 seconds are spent in apply_solver, which is where the subprocess command that calls Gurobi is executed. These results are an indication a persistent solver interface would be useful for this application. The persistent solver interfaces will be discussed in section 9.3.

9.2 Improving Model Construction Performance with LinearExpression

In this section, we discuss using the LinearExpression class directly in order to improve model construction time. Although operator overloading is a very convenient way to construct expressions, it can be computationally expensive. Alternatively, Pyomo supports creating linear expressions using native Python lists, which can be significantly faster than the operator overloading approach. The constructor of the LinearExpression class takes three keyword arguments:

- constant: a constant term
- linear_vars: a list of Pyomo variables appearing in the linear expression
- linear_coefs: a list with coefficients for each of the variables in the linear_vars list

The following example compares creating a linear expression using operator overloading and the LinearExpression constructor. In order to compare timing, we create the expression 100,000 times.

```
import pyomo.environ as pyo
from pyomo.common.timing import TicTocTimer
from pyomo.core.expr.numeric_expr import LinearExpression

N1 = 10
N2 = 100000
```

```
m = pyo.ConcreteModel()
m.x = pyo.Var(list(range(N1)))

timer = TicTocTimer()
timer.tic()

for i in range(N2):
    e = sum(i*m.x[i] for i in range(N1))
timer.toc('created expression with sum function')

for i in range(N2):
    coefs = [i for i in range(N1)]
    lin_vars = [m.x[i] for i in range(N1)]
    e = LinearExpression(constant=0, linear_coefs=coefs, \
        linear_vars=lin_vars)
timer.toc('created expression with LinearExpression constructor')
```

The output is shown next.

```
[ 0.00] Resetting the tic/toc delta timer
[+ 3.52] created expression with sum function
[+ 0.52] created expression with LinearExpression constructor
```

Although using the `LinearExpression` is less clear and less concise, it is over 6 times faster in this example. Note the performance improvement depends heavily on the number of terms in the expression.

9.3 Repeated Solves with Persistent Solvers

Each time a Pyomo model is solved using the standard solver interfaces, the generalized modeling components defining the model must be translated into some input form recognized by the solver in use. For cases where the solver finishes relatively quickly, this model translation may introduce significant timing overhead. Persistent solver interfaces provide a mechanism for reducing overall translation time for models that are solved repeatedly with incremental changes. This is made possible by exposing functionality allowing users to efficiently notify the solver of these incremental changes to the Pyomo model, after an initial step where the full model is translated.

When a persistent interface is used, additional care must be taken so that the Pyomo model and its translated solver representation are kept in sync. This is a manual process that must be done by the user. However, with this tradeoff, significant performance improvements are possible for many common optimization approaches.

In Section 9.3.1, we briefly discuss when persistent solvers are most useful. In Sections 9.3.2 – 9.3.4, we describe how to use a persistent solver interface. In Section 9.3.5, we re-implement the parametric sweep example from Section 9.1.3 using a persistent solver interface.

9.3.1 When to Use a Persistent Solver

Persistent solver interfaces are designed to be used when repeatedly solving the same model with minor changes. They are most useful when the solver time is not significantly longer than the time needed for Pyomo to translate the model to the solver's input format. Of course, persistent solvers can be used no matter how long it takes to solve the problem. However, if the time spent solving the problem is significantly longer than the time needed for translation, then little speedup will be observed, but the complexity of the code may have increased. Linear programs are excellent candidates for use with persistent solver interfaces because most linear programs can be solved very efficiently. On the other hand, many mixed-integer programs are difficult to solve and are poor candidates for the persistent solver interfaces. Ultimately, it is necessary to use the profilers discussed in Section 9.1 to determine how beneficial a persistent solver interface will be for any particular application.

9.3.2 Basic Usage

The first step in using a persistent solver is to create a Pyomo model as usual.

```
import pyomo.environ as pyo
m = pyo.ConcreteModel()
m.x = pyo.Var()
m.y = pyo.Var()
m.obj = pyo.Objective(expr=m.x**2 + m.y**2)
m.c = pyo.Constraint(expr=m.y >= -2*m.x + 5)
```

You can create an instance of a persistent solver through the `SolverFactory`.

```
opt = pyo.SolverFactory('gurobi_persistent')
```

This returns an instance of the `GurobiPersistent` class. Now we need to tell the solver about our model.

```
opt.set_instance(m)
```

This will create a gurobipy Model object and include the appropriate variables and constraints. We can now solve the model.

```
results = opt.solve()
```

Note that the model should not be passed into the `solve` method, as is done with most solver interfaces. We can also add or remove variables, constraints, or blocks and set objectives. For example,

```
m.c2 = pyo.Constraint(expr=m.y >= m.x)
opt.add_constraint(m.c2)
```

This tells the solver to add one new constraint but otherwise leave the model unchanged. We can now resolve the model.

```
results = opt.solve()
```

To remove a component, simply call the corresponding remove method.

```
opt.remove_constraint(m.c2)
del m.c2
results = opt.solve()
```

If a Pyomo component is replaced with another component with the same name, the first component must be removed from the solver. Otherwise, the solver will have multiple components. For example, the following code will run without error, but the solver will have an extra constraint. The solver will have both $y \geq -2*x+5$ and $y \leq x$, which is not what was intended!

```
m = pyo.ConcreteModel()
m.x = pyo.Var()
m.y = pyo.Var()
m.c = pyo.Constraint(expr=m.y >= -2*m.x + 5)
opt = pyo.SolverFactory('gurobi_persistent')
opt.set_instance(m)
# WRONG:
del m.c
m.c = pyo.Constraint(expr=m.y <= m.x)
opt.add_constraint(m.c)
```

The correct way to do this is:

```
m = pyo.ConcreteModel()
m.x = pyo.Var()
m.y = pyo.Var()
m.c = pyo.Constraint(expr=m.y >= -2*m.x + 5)
opt = pyo.SolverFactory('gurobi_persistent')
opt.set_instance(m)
# Correct:
opt.remove_constraint(m.c)
del m.c
m.c = pyo.Constraint(expr=m.y <= m.x)
opt.add_constraint(m.c)
```

In most cases, the only way to modify a component is to remove it from the solver instance, modify it with Pyomo, and then add it back to the solver instance. The only exception is with variables. Variables may be modified and then updated with the solver:

```
m = pyo.ConcreteModel()
m.x = pyo.Var()
m.y = pyo.Var()
m.obj = pyo.Objective(expr=m.x**2 + m.y**2)
m.c = pyo.Constraint(expr=m.y >= -2*m.x + 5)
opt = pyo.SolverFactory('gurobi_persistent')
opt.set_instance(m)
m.x.setlb(1.0)
opt.update_var(m.x)
```

In short, any time the Pyomo model is changed, the persistent solver interface must be notified and kept in sync. Table 9.1 presents the appropriate methods to use for various Pyomo model modifications. Note that when mutable parameters or

Table 9.1: Persistent Solver Interface Methods

Add constraint	`opt.add_constraint()`
Add variable	`opt.add_var()`
Add block	`opt.add_block()`
Set objective	`opt.set_objective()`
Remove constraint	`opt.remove_constraint()`
Remove variable	`opt.remove_var()`
Remove block	`opt.remove_block()`
Modify variable	`opt.update_var()`

Modify mutable parameter	`opt.remove_constraint()`
	`m.param.value = val`
	`opt.add_constraint()`

Modify `Expression`	`opt.remove_constraint()`
	`m.expr += val`
	`opt.add_constraint()`

named `Expressions` are modified, all constraints utilizing the modified parameters/expressions must be updated (removed and re-added).

9.3.3 Working with Indexed Variables and Constraints

The examples in section 9.3.2 all used scalar variables and constraints; in order to use indexed variables and/or constraints, the code must be slightly adapted:

```
m.v = pyo.Var([0, 1, 2])
m.c2 = pyo.Constraint([0, 1, 2])
for i in range(3):
    m.c2[i] = m.v[i] == i
for v in m.v.values():
    opt.add_var(v)
for c in m.c2.values():
    opt.add_constraint(c)
```

This must be done when removing indexed variables and constraints, too. Note that the `is_indexed` method can be used to automate the process.

9.3.4 Additional Performance

In order to get the best performance out of the persistent solvers, use the `save_results` argument:

```
m = pyo.ConcreteModel()
m.x = pyo.Var()
m.y = pyo.Var()
m.obj = pyo.Objective(expr=m.x**2 + m.y**2)
m.c = pyo.Constraint(expr=m.y >= -2*m.x + 5)
opt = pyo.SolverFactory('gurobi_persistent')
opt.set_instance(m)
results = opt.solve(save_results=False)
```

Note that if the save_results flag is set to False, then the following is not supported.

```
results = opt.solve(save_results=False, load_solutions=False)
if results.solver.termination_condition == \
    pyo.TerminationCondition.optimal:
  try:
      m.solutions.load_from(results)
  except AttributeError:
      print('AttributeError was raised')
```

```
AttributeError was raised
```

However, the following will work:

```
results = opt.solve(save_results=False, load_solutions=False)
if results.solver.termination_condition == \
    pyo.TerminationCondition.optimal:
  opt.load_vars()
```

Additionally, a subset of variable values may be loaded back into the model:

```
results = opt.solve(save_results=False, load_solutions=False)
if results.solver.termination_condition == \
    pyo.TerminationCondition.optimal:
  opt.load_vars([m.x])
```

9.3.5 Example

In this section, we re-implement the parameter sweep example from Section 9.1.3 using a persistent solver interface. The following is a function which performs the parameter sweep with a persistent solver interface.

```
def solve_parametric_persistent():
  m = create_warehouse_model(num_locations=50, num_customers=50)
  opt = pyo.SolverFactory('gurobi_persistent')
  opt.set_instance(m)
  p_values = list(range(1, 31))
  obj_values = list()
  for p in p_values:
      m.P.value = p
      opt.remove_constraint(m.num_warehouses)
      opt.add_constraint(m.num_warehouses)
```

```
res = opt.solve(save_results=False)
assert_optimal_termination(res)
obj_values.append(res.problem.lower_bound)
```

There are a few additional lines of code compared to the function from Section 9.1.3. Specifically, we have added calls to the `set_instance`, `remove_constraint`, and `add_constraint` methods on the solver interface.

The following code block calls the above function and reports the execution time.

```
timer.tic()
solve_parametric_persistent()
timer.toc('Finished parameter sweep with persistent interface')
```

The output is here.

```
[ 22.79] Resetting the tic/toc delta timer
[+ 0.91] finished parameter sweep with persistent interface
```

Note this function was approximately 8 times faster than the non-persistent counterpart (0.91 seconds vs 7.28 seconds). Recall that the performance depends heavily on the application as discussed in Section 9.3.1.

9.4 Sparse Index Sets

Often, a model makes use of only a portion of the cross product of multiple index sets. In particular, sometimes the members in one dimension depend on the value in another, leading to the slang "jagged sets."

For example, a modeler may want to have a constraint to hold for

$$i \in \mathscr{I}, k \in \mathscr{K}, v \in \mathscr{V}_k.$$

There are many ways to accomplish this, but one good way is to create a set of tuples composed of all of valid "model.k, model.V[k]" pairs. This is illustrated in the following example where the jagged set KV is used.

```
import pyomo.environ as pyo

model = pyo.AbstractModel()

model.I = pyo.Set()
model.K = pyo.Set()
model.V = pyo.Set(model.K)

def kv_init(m):
    return ((k,v) for k in m.K for v in m.V[k])
model.KV = pyo.Set(dimen=2, initialize=kv_init)

model.a = pyo.Param(model.I, model.K)

model.y = pyo.Var(model.I)
model.x = pyo.Var(model.I, model.KV)
```

```
# include a constraint that looks like this:
# x[i,k,v] <= a[i,k]*y[i], for i in I, k in K, v in V[k]

def c1Rule(m,i,k,v):
  return m.x[i,k,v] <= m.a[i,k]*m.y[i]
model.c1 = pyo.Constraint(model.I, model.KV, rule=c1Rule)
```

An alternative strategy would be to declare the constraint to be indexed by sets I, K, and V and then use `pyo.Constraint.Skip()` to pass over the indices that are not present. However, in higher dimension, or with large sets this can cause significant performance degradation.

Chapter 10
Abstract Models and Their Solution

Abstract This chapter describes how to declare and use an `AbstractModel` and data command files to initilize abstract models. Finally, this chapter describes the `pyomo` command, which makes it particularly easy to solve an abstract model using data command files. Although concrete and abstract models provide similar functionality, abstract models make a strong seperation of model formulation and model data, which is conceptually nice and practically useful in some contexts.

10.1 Overview

In many of the examples in this book, we use a function that takes in data and returns a `ConcreteModel`. This facilitates separating the concepts of model and data. The `AbstractModel` class in Pyomo provides this separation by populating the model with data only after the abstract model object has been created.

10.1.1 Abstract and Concrete Models

Pyomo supports two strategies for model declaration: concrete models, which immediately construct model components, and abstract models, which defer component construction. Abstract models reflect the structure of many mathematical optimization formulations. For example, the formulation of the warehouse location problem (WL) on page 26 is written in a general manner describing a class of optimization problems. However, we cannot *solve* this problem because the actual data for the problem (N, M, d, and P) have not been specified. A solver must be given a specific instance of a problem (with the data specified).

In Pyomo, an abstract model is declared first, and component construction is delayed until the data is loaded and Pyomo creates the model instance. This modeling approach is illustrated in the top pane of Figure 10.1. An `AbstractModel` object

M. L. Bynum et al., *Pyomo — Optimization Modeling in Python*, Springer Optimization and Its Applications 67, https://doi.org/10.1007/978-3-030-68928-5_10

is created, and then the data for a particular problem is given to Pyomo, and then Pyomo performs the construction process in order to create an *instance* of the model with all the variables, constraint expressions, and objective expressions that can be sent to a solver. This requires a two-pass approach where the model is declared in the first pass, and subsequently the model is constructed using data values specified separately. To support delayed construction, the model must be defined using construction rules.

By contrast, concrete models support a programmatic style where the model instance is created immediately; model components are constructed and initialized on the first pass as Python executes the model script. This modeling approach is illustrated in the bottom pane of Figure 10.1. A `ConcreteModel` object is created, and the data needs to be present *before* each component is declared. As Python executes your model script, the particular model instance and its components are created immediately as Python encounters the component declaration. Once the execution through the Python file is completed the model is ready to be sent to the solver (i.e., a single pass). At this point, the `ConcreteModel` *is* the specific instance.

NOTE: Construction rules can be still be used with concrete models (the rules are immediately fired as they are encountered in the model's Python file).

AbstractModel construction process

ConcreteModel construction process

Fig. 10.1: This figure describes the construction process for both abstract and concrete models. The top pane describes the declarative style used for abstract models. The `AbstractModel` is first created. Then, given a particular realization of the data, Pyomo performs the construction process in order to create an *instance* of the model that can be sent to the solver. The bottom pane describes the programmatic style used for concrete models. As Python executes the model script, the component objects are constructed immediately using data previously declared. The particular model *instance* is ready to be sent to the solver once the first pass is complete.

The choice of which model object to use (AbstractModel or ConcreteModel) is largely a matter of taste and preference. The biggest difference between the use of AbstractModel or ConcreteModel relates to the specification of data. When using an AbstractModel, the names and structure of the data used in a model are declared prior to construction (but not the values themselves). This means Pyomo is aware of the existence of the sets and parameters that *will* be encountered when constructing the instance. More importantly, Pyomo knows the associated names and types of these quantities and something about the relationships among them (e.g., the variable y is indexed by set N). This allows the user to specify the data using any number of Pyomo supported data formats while referring to these quantities by name. Pyomo includes many options for supplying data to an abstract model, including a *data command file* to specify values for set and parameter data. The syntax of Pyomo's data command files is very similar to the data command syntax supported by AMPL [2].

Concrete models facilitate more straightforward use of native Python data types when creating a model instance. Therefore, if you are more comfortable building models in a procedural programming environment (like Python or MATLAB), or if your application requires a more extensive workflow than that supported by the pyomo command, then a ConcreteModel is more appropriate. This is especially true if your data can be easily loaded into Python through other Python packages (e.g., pandas).

NOTE: In general, an AbstractModel is more straightforward for users that are unfamiliar with Python or prefer to work in some more traditional AML environments. A ConcreteModel often requires Python coding on the part of the user to load the data (e.g., using an existing Python package for the raw data format) and apply it to the model, but offers transparent control over execution order.

10.1.2 An Abstract Formulation of Model (H)

Consider model (H) (see page 19), which is reproduced here for convenience:

$$\max_{x} \sum_{i \in \mathscr{A}} h_i \cdot \left(x_i - (x_i/d_i)^2 \right) \quad (H)$$
$$\text{s.t.} \sum_{i \in \mathscr{A}} c_i x_i \leq b$$
$$0 \leq x_i \leq u_i, \ i \in \mathscr{A}$$

Since model (H) is an abstract model, a natural way of expressing this model in Pyomo is with Pyomo's AbstractModel class.

Consider the following Pyomo model for this problem:

```python
# AbstractH.py - Implement model (H)
import pyomo.environ as pyo

model = pyo.AbstractModel(name="(H)")

model.A = pyo.Set()

model.h = pyo.Param(model.A)
model.d = pyo.Param(model.A)
model.c = pyo.Param(model.A)
model.b = pyo.Param()
model.u = pyo.Param(model.A)

def xbounds_rule(model, i):
    return (0, model.u[i])
model.x = pyo.Var(model.A, bounds=xbounds_rule)

def obj_rule(model):
    return sum(model.h[i] * \
            (model.x[i] - (model.x[i]/model.d[i])**2) \
            for i in model.A)
model.z = pyo.Objective(rule=obj_rule, sense=pyo.maximize)

def budget_rule(model):
    return sum(model.c[i]*model.x[i] for i in model.A) <= model.b
model.budgetconstr = pyo.Constraint(rule=budget_rule)
```

Given particular data for the parameters in this model, then one might be interested in finding an optimal assignment of values to `model.x`. There are a variety of ways to provide data to Pyomo for an abstract model. Here is data (saved in `AbstractH.dat`) defining a suitable happiness objective for one of the authors of this book:

```
# Pyomo data file for AbstractH.py
set A := I_C_Scoops Peanuts ;
param h := I_C_Scoops 1 Peanuts 0.1 ;
param d :=
  I_C_Scoops 5
  Peanuts 27 ;
param c := I_C_Scoops 3.14 Peanuts 0.2718 ;
param b := 12 ;
param u := I_C_Scoops 100 Peanuts 40.6 ;
```

This is a Pyomo data file, which includes `set` and `param` commands closely resembling AMPL data commands. Description of these data commands is given in Section 10.3.

The following lines can be used to optimize an abstract model, by adding them to the Python file defining the model:

```
opt = pyo.SolverFactory('glpk')

instance = model.create_instance("AbstractH.dat")
results = opt.solve(instance) # solves and updates instance

instance.display()
```

Alternatively, we can also solve the model using the `pyomo` command as described in Section 10.2

10.1.3 An Abstract Model for the Warehouse Location Problem

The warehouse location problem (see Section 3.2) can be represented as an abstract model as follows:

```
 1  # wl_abstract.py: AbstractModel version of warehouse \
        location determination problem
 2  import pyomo.environ as pyo
 3
 4  model = pyo.AbstractModel(name="(WL)")
 5  model.N = pyo.Set()
 6  model.M = pyo.Set()
 7  model.d = pyo.Param(model.N,model.M)
 8  model.P = pyo.Param()
 9  model.x = pyo.Var(model.N, model.M, bounds=(0,1))
10  model.y = pyo.Var(model.N, within=pyo.Binary)
11
12  def obj_rule(model):
13      return sum(model.d[n,m]*model.x[n,m] for n in \
            model.N for m in model.M)
14  model.obj = pyo.Objective(rule=obj_rule)
15
16  def one_per_cust_rule(model, m):
17      return sum(model.x[n,m] for n in model.N) == 1
18  model.one_per_cust = pyo.Constraint(model.M, \
        rule=one_per_cust_rule)
19
20  def warehouse_active_rule(model, n, m):
21      return model.x[n,m] <= model.y[n]
22  model.warehouse_active = pyo.Constraint(model.N, \
        model.M, rule=warehouse_active_rule)
23
24  def num_warehouses_rule(model):
25      return sum(model.y[n] for n in model.N) <= model.P
26  model.num_warehouses = \
        pyo.Constraint(rule=num_warehouses_rule)
```

The sets are declared as Pyomo `Set` components and the parameter data is declared as Pyomo `Param` components with no indication as to how the data will be supplied. Pyomo is informed the `Param` and `Var` components will be indexed by sets, but the contents of those sets have not been declared.

The objective construction rule is defined on as the Python function `obj_rule` and then `model.obj` is declared to be Pyomo `Objective` component. Since this is an abstract model, the objective rule `obj_rule` is not yet called. At this point, Pyomo knows what rule to call to construct the objective component, but it has not called the constructor because this is an abstract model. Constraint rules and constraint components are declared in a similar manner.

Data can be expressed in several different formats. For example, the following Pyomo data file can be used:

```
# wl_data.dat: Pyomo format data file for the warehouse \
    location problem

set N := Harlingen Memphis Ashland ;
set M := NYC LA Chicago Houston;

param d :=
    Harlingen NYC 1956
    Harlingen LA  1606
    Harlingen Chicago  1410
    Harlingen Houston  330
    Memphis NYC  1096
    Memphis LA  1792
    Memphis Chicago  531
    Memphis Houston  567
    Ashland NYC  485
    Ashland LA  2322
    Ashland Chicago  324
    Ashland Houston  1236
;

param P := 2 ;
```

The script can be executed with the `python` command, but this action would not actually *do* anything. This script declares the model, but it does not define the model data or create the problem instance for the solver. The action of applying a data file to this abstract model can be scripted explicitly in Python code, or it can be done using the `pyomo` command. For example:

```
pyomo solve --solver=glpk wl_abstract.py wl_data.dat
```

The `--summary` flag can be used to provide more detailed output about the solution.

When `pyomo` runs, it executes `wl_abstract.py` to create an `AbstractModel` with the name `model`. This model object contains the Pyomo modeling components that have been declared. Then `pyomo` reads the data file `wl_data.dat` and applies this data to the `Set` and `Param` components in the same order that the components were declared in the model. Next, the `pyomo` command constructs all of the remaining components in declaration order: the variables, the objective, and the constraints. After the model is constructed, pyomo calls the solver to find the solution.

Abstract models can be used in scripts, but a concrete instance must be created from the AbstractModel object using the create_instance method. The following example takes an AbstractModel, constructs the instance using a data file called wl_data.dat, solves the instance, and prints some results:

```
instance = model.create_instance('wl_data.dat')
solver = pyo.SolverFactory('glpk')
solver.solve(instance)
instance.y.pprint()
```

If you don't want to write your own script, but instead want to use the pyomo command, then you should not add these lines.

10.2 The pyomo Command

The Pyomo software distribution includes a pre-defined execution script, the pyomo command, that includes a variety of subcommands supporting use of Pyomo. The following subcommands are supported in Pyomo 6.0:

check
> This subcommand checks a model for errors. This is particularly useful for evaluating the logic of rules in abstract models.

convert
> This subcommand is used to convert a Pyomo model into another format, such as an lp or nl file.

help
> Print information about the configuration and installation of Pyomo. For example, the -s option provides information about available solvers:
> ```
> pyomo help -s
> ```

run
> Execute a command from the Pyomo bin (or Scripts) directory. For example, this provides a handy mechanism for launching Python with Pyomo installed:
> ```
> pyomo run python
> ```

solve
> Construct and optimize a model.

test-solvers
> Execute a variety of tests to verify solver capabilities.

The following sections illustrate the use of the check, convert, help and solve subcommands, which can be customized with a variety of options.

10.2.1 The `help` Subcommand

The `help` subcommand prints information about Pyomo's capabilities, including information about installed plugins as well as available solvers. The −h option prints information about the different information available, including:

−−checkers
> This prints the model checkers that are installed with Pyomo.

−−commands, −c
> This prints the commands installed with Pyomo. Although much of Pyomo's functionality can be accessed through the `pyomo` command, some functionality is developed separately.

−−components
> This prints the modeling components and virtual sets available in Pyomo.

−−data−managers, −d
> This prints the data interfaces supported by the `DataPortal` class. Data can also be imported through these interfaces using Pyomo data files.

−−info, −i
> This prints information about the user's PATH environment and Python installation. This command helps diagnose issues with the execution of Pyomo.

−−solvers, −s
> This prints information about solvers and solver managers that can be used the optimize Pyomo models. Note that information about NEOS solvers will be included if Pyomo can connect to the NEOS server.

−−transformations, −t
> This prints the model transformations supported by Pyomo.

−−writers, −w
> This prints the model writers supported by Pyomo. Specifically, this summarizes the different file formats a Pyomo model can be converted to.

So, for example, to see a list of available transformations, use the command:

```
pyomo help --transformations
```

10.2.2 The `solve` Subcommand

The `pyomo solve` subcommand automatically executes a Pyomo model as follows:

1. Construct a model.

2. Read the instance data (if applicable).

3. Generate a model instance (if the model is abstract).

4. Apply simple preprocessors to the model instance.

5. Apply a solver to the model instance.

6. Load the results into the model instance.

7. Display the solver results.

For example, the following command solves the warehouse location problem defined in `wl_abstract.py` using the LP solver `glpk` using data from the file `wl_data.dat`:

```
pyomo solve --solver=glpk wl_abstract.py wl_data.dat
```

The model construction step requires a Pyomo A *Pyomo model file*, which is a Python file defining a Pyomo model object. Thus, the `solve` subcommand can be viewed as a generic script for analyzing a model defined by a Pyomo model file. The `solve` subcommand has a variety of optional command-line arguments to customize the optimization process; documentation of the various available options is available by specifying the `--help` option.

However, the `solve` subcommand can also be executed with a YAML or JSON configuration file[1], which eliminates the need to specify command-line options. Consider the following configuration file:

```
# concrete1.yaml
model:
   filename: concrete1.py
solvers:
 - solver name: glpk
```

This configuration file can be used to configure the executions of the `pyomo` subcommand as follows:

```
pyomo solve concrete1.yaml
```

This configuration file defines the same logic as the first command in the previous paragraph, and the following configuration file defines the same logic as the second command:

```
# abstract5.yaml
model:
   filename: abstract5.py
data:
   files:
    - abstract5.dat
solvers:
 - solver name: glpk
```

No command-line options are required when using a configuration file, because all command-line options have corresponding elements in a configuration file. Furthermore, there are configuration options that can only be expressed in a configuration file. A template configuration file can be generated with the `--generate-config-template` option .

---

[1] YAML and JSON are data serialization standards. JSON is supported natively in Python, and information about JSON is available at `www.json.org`. YAML configuration files are supported if the `PyYAML` package is installed, and information about YAML is available at `www.yaml.org`.

Option	Description
`-c, --catch-errors`	Trigger failures for exceptions and print the program stack.
`--json`	Store results in JSON format.
`-k, --keepfiles`	Keep temporary files.
`-l, --log`	Print the solver logfile after performing optimization.
`--logfile FILE`	Redirect output to the specified file.
`--logging LEVEL`	Specify the logging level: quiet, warning, info, verbose, debug.
`--model-name NAME`	The name of the model object created in the specified Pyomo module.
`--path PATH`	Give a path used to find Pyomo Python files.
`--report-timing`	Report various timing statistics during model construction.
`--results-format FORMAT`	Specify the results format: json or yaml.
`--save-results FILE`	The filename to which the results are saved.
`--show-results`	Print the results object after optimization.
`--solver SOLVER`	Specify the solver name.
`--solver-executable FILE`	The executable used by the solver interface.
`--solver-io FORMAT`	The type of IO used to execute the solver. Different solvers support different types of IO, but the following are common options: lp - generate LP files, nl - generate NL files, python - direct Python interface.
`--stream-output`	Stream the solver output to provide information about the solver's progress.
`--solver-options STRING`	String describing solver options.
`--solver-suffix SUFFIXES`	Solution suffixes that will be extracted by the solver (e.g., rc, dual, or slack).
`--summary`	Summarize the final solution after performing optimization.
`--symbolic-solver-labels`	When interfacing with the solver, use symbol names derived from the model. For example, "my_special_variable[1_2_3]" instead of "v1". When using the ASL solvers, this option generates corresponding .row (constraints) and .col (variables) files.
`--tempdir TEMPDIR`	Specify the directory where temporary files are generated.

Table 10.1: Commonly used options for the `pyomo solve` subcommand.

The `--help` and `--generate-config-template` options for the `solve` subcommand require the `--solver` option. These two options provide solver-specific summaries respectively for command-line options and configuration files. For example, you could execute the following command to get command-line options suitable for the `glpk` solver:

```
pyomo solve --solver=glpk --help
```

Table 10.1 summarizes key options for the `solve` subcommand that are commonly used.

10.2.2.1 Specifying the Model Object

A model file can execute an arbitrary Python script, but the expectation of the pyomo solve command is that it generates an object that contains the Pyomo model. Within the solve subcommand, a model file is executed with a Python import command, and thus it is interpreted like any other Python file.

In the simplest case, a Pyomo model file contains Python commands to create a model object stored in the model variable. For example, consider the following simple LP:

```python
# abstract5.py
import pyomo.environ as pyo

model = pyo.AbstractModel()

model.N = pyo.Set()
model.M = pyo.Set()
model.c = pyo.Param(model.N)
model.a = pyo.Param(model.N, model.M)
model.b = pyo.Param(model.M)

model.x = pyo.Var(model.N, within=pyo.NonNegativeReals)

def obj_rule(model):
    return sum(model.c[i]*model.x[i] for i in model.N)
model.obj = pyo.Objective(rule=obj_rule)

def con_rule(model, m):
    return sum(model.a[i,m]*model.x[i] for i in model.N) \
               >= model.b[m]
model.con = pyo.Constraint(model.M, rule=con_rule)
```

This is an abstract Pyomo model stored in the model variable.

If a user defines their model with a different variable name, then the --model-name option can be used to direct Pyomo to select the specified name. For example, we can adapt the previous example to store the model in Model:

```python
# abstract6.py
import pyomo.environ as pyo

Model = pyo.AbstractModel()

Model.N = pyo.Set()
Model.M = pyo.Set()
Model.c = pyo.Param(Model.N)
Model.a = pyo.Param(Model.N, Model.M)
Model.b = pyo.Param(Model.M)

Model.x = pyo.Var(Model.N, within=pyo.NonNegativeReals)

def obj_rule(Model):
    return sum(Model.c[i]*Model.x[i] for i in Model.N)
Model.obj = pyo.Objective(rule=obj_rule)
```

```
def con_rule(Model, m):
    return sum(Model.a[i,m]*Model.x[i] for i in Model.N) \
                >= Model.b[m]
Model.con = pyo.Constraint(Model.M, rule=con_rule)
```

This model can be optimized with the following command:

```
pyomo solve --solver=glpk --model-name=Model \
            abstract6.py abstract6.dat
```

Aside from supporting greater flexibility for the user, this option allows users to define multiple models in a Pyomo model file and then select the model to be optimized when the `solve` subcommand is executed.

10.2.2.2 Selecting Data with Namespaces

Section 10.3.4 introduces the `namespace` command in Pyomo data files. This command is used to define blocks of data commands that are integrated optionally into a model. The `solve` subcommand provides the `--namespace` option to specify one or more namespaces used to construct an instance of an abstract model; the `--ns` is a shorter alias for this option. For example, the command

```
pyomo solve --solver=glpk --namespace=data1 abstract5.py \
                abstract5-ns1.dat
```

creates and optimizes the abstract model in `abstract5.py` using the following data commands:

```
namespace data1 {
    set N := 1 2 ;

    set M := 1 2 ;

    param c :=
    1 1
    2 2 ;

    param a :=
    1 1 3
    2 1 4
    1 2 2
    2 2 5 ;

    param b :=
    1 1
    2 2 ;
}

namespace data2 {
    set N := 3 4 ;

    set M := 5 6 ;
```

```
    param c :=
    3 10
    4 20 ;

    param a :=
    3 5 3
    4 5 4
    3 6 2
    4 6 5 ;

    param b :=
    5 1
    6 2 ;
}
```

This command specifies the `data1` namespace, which has an optimal solution of 0.8. Similarly, the command

```
pyomo solve --solver=glpk --namespace=data2 abstract5.py \
                    abstract5-ns1.dat
```

creates and optimizes the same model using the `data2` namespace, which has an optimal solution of 8. A different index set is used in the `data2` data, as well as different objective coefficients.

The previous example illustrates how namespaces allow the user to specify different data sets within a single data command file. Note that a model can be constructed from data commands using multiple namespaces, including data not in a namespace. Consider the following data commands:

```
set N := 1 2;

namespace c1 {
    param c :=
    1 1
    2 2 ;
}

namespace c2 {
    param c :=
    1 10
    2 20 ;
}

namespace data1 {
    set M := 1 2 ;

    param a :=
    1 1 3
    2 1 4
    1 2 2
    2 2 5 ;

    param b :=
```

```
    1 1
    2 2 ;
}

namespace data2 {
    set M := 5 6 ;

    param a :=
    1 5 3
    2 5 4
    1 6 2
    2 6 5 ;

    param b :=
    5 1
    6 2 ;
}
```

This includes four namespaces and data commands outside of a namespace. The command

```
pyomo solve --solver=glpk --namespace=c1 --namespace=data2 \
            abstract5.py abstract5-ns2.dat
```

creates and optimizes the abstract modeling in abstract5.py using data commands from the c1 and data2 namespaces, as well as the data command for N, which is outside of any namespace. Note that if multiple namespaces contain data commands for the same component, then the component is initialized with the data from first namespace containing the corresponding data command. If there is not a namespace containing a corresponding data command, then the data commands outside of namespaces are used to initialize the component.

10.2.2.3 Customizing Pyomo's Workflow

The different steps that are executed by the solve subcommand represent a generic workflow for model construction and optimization. This workflow can be customized using a variety of callback functions that are defined within a Pyomo model file. These callback functions allow the user to define additional analysis steps, as well as replace some of the default steps in the workflow.

Function	Description
pyomo_preprocess	Perform a preprocessing step before model construction
pyomo_create_model	Construct and return a model object
pyomo_create_modeldata	Construct and return a DataPortal object
pyomo_print_model	Output model object information
pyomo_modify_instance	Modify the model instance
pyomo_print_instance	Output model instance information
pyomo_save_instance	Save the model instance
pyomo_print_results	Print the optimization results
pyomo_save_results	Save the optimization results
pyomo_postprocess	Perform a postprocessing step after optimization

Table 10.2: Callback functions that can be used in a Pyomo model file to customize the workflow in the pyomo solve subcommand.

Table 10.2 summarizes the pyomo command callback functions and the functionality that they support. Each callback function takes one or more keyword arguments in the form keyword=value. For example, the pyomo_print_results callback function takes three arguments: options, instance, and results.

```
def pyomo_print_results(options=None, instance=None,
                        results=None):
    print(results)
```

There are several standard arguments for the callback functions described in Table 10.2. The options argument is an enhanced Python dictionary containing the command-line options sent to the solve subcommand. The model argument is the Pyomo model object, and the instance argument is the model instance constructed from this model. In the case where the user defines a model using ConcreteModel, then the model and instance arguments are the same object. Other arguments are described with their associated callback functions.

pyomo_preprocess

This callback function is executed before model construction to perform preprocessing steps. This function has one argument: options. For example, the following callback function simply prints the command-line options:

```
def pyomo_preprocess(options=None):
    print("Here are the options that were provided:")
    if options is not None:
        options.display()
```

pyomo_create_model

This callback function is used to construct a model. This function has two arguments: `options` and `model_options`. The latter argument contains the options for constructing the model, which are specified with the `--model-options` command-line option. The return value of this function must be the model object created, which may be either an abstract or concrete model. For example, the following callback function creates a model by importing the `abstract6.py` file and then returning the `Model` object:

```
def pyomo_create_model(options=None, model_options=None):
    sys.path.append(abspath(dirname(__file__)))
    abstract6 = __import__('abstract6')
    sys.path.remove(abspath(dirname(__file__)))
    return abstract6.Model
```

pyomo_create_modeldata

This callback function creates a model data object used to create a model instance. Model data objects are useful in contexts where a set of different data sources need to be specified for model constructions. This function has two arguments: `options` and `model`. The return value must be a `DataPortal` object. For example, the following callback function creates a `DataPortal` object from the file `abstract6.dat`:

```
def pyomo_create_dataportal(options=None, model=None):
    data = pyo.DataPortal(model=model)
    data.load(filename='abstract6.dat')
    return data
```

pyomo_print_model

This callback function prints an abstract model before a model instance is created. This function has two arguments: `options` and `model`. The following example calls the `pprint` method to print detailed information about an abstract model:

```
def pyomo_print_model(options=None, model=None):
    if options['runtime']['logging']:
        model.pprint()
```

pyomo_modify_instance

This callback function modifies the model instance after it has been constructed. This function has three arguments: `options`, `model`, and `instance`. The following callback fixes a variable after the model is constructed:

```
def pyomo_modify_instance(options=None, model=None,
                          instance=None):
    instance.x[1].value = 0.0
    instance.x[1].fixed = True
```

pyomo_print_instance

This callback function prints the Pyomo model instance. This function is used to print the concrete model instance rather than the abstract model. This function has two arguments: options and instance. The following example calls the pprint method to print detailed information about a model instance:

```
def pyomo_print_instance(options=None, instance=None):
    if options['runtime']['logging']:
        instance.pprint()
```

pyomo_save_instance

This callback function saves the Pyomo model instance. This function has two arguments: options and instance. Note that Pyomo does not specify how the model is saved. However, a convenient mechanism would be to use Python's pickle mechanism:

```
def pyomo_save_instance(options=None, instance=None):
    OUTPUT = open('abstract7.pyomo','w')
    OUTPUT.write(str(pickle.dumps(instance)))
    OUTPUT.close()
```

pyomo_print_results

This callback function prints the results generated from optimization. This function has three arguments: options, instance, and results. The results object supports a generic summary of optimization solutions, solver statistics, etc. in both the JSON or YAML formats. Thus, this callback function can simply print this data:

```
def pyomo_print_results(options=None, instance=None,
                        results=None):
    print(results)
```

However, the solve subcommand includes the --print-results option, which performs this operation. More generally, this callback function is included to allow users to provide problem-specific summaries of their optimization results.

pyomo_save_results

This callback function is used to save the results generated from optimization. This function has three arguments: `options`, `instance`, and `results`. This callback function can simply print the results to a file:

```
def pyomo_save_results(options=None, instance=None,
                       results=None):
    OUTPUT = open('abstract7.results','w')
    OUTPUT.write(str(results))
    OUTPUT.close()
```

The `solve` subcommand includes the `--save-results` option, which performs this operation. More generally, this callback function is included to allow users to save problem-specific summary of their optimization results.

pyomo_postprocess

This callback function is executed after optimization to perform postprocessing steps. This function has three arguments: `options`, `instance`, and `results`. For example, the following function prints a simple summary of the optimization results:

```
def pyomo_postprocess(options=None, instance=None,
                      results=None):
    instance.solutions.load_from(results, \
            allow_consistent_values_for_fixed_vars=True)
    print("Solution value "+str(pyo.value(instance.obj)))
```

10.2.2.4 Customizing Solver Behavior

The generic workflow supported by the `solve` subcommand includes the execution of a solver to optimize (or otherwise analyze) a model. A variety of command-line options are used to control solver behavior. The `--solver` option is used to specify the name of the solver constructed. This option can specify two classes of solvers: the names of command-line executables on the user's path, and predefined solver interfaces.

Command-line executables are assumed to perform I/O using NL files. Thus, command-line executables can be optimized with any solver executable built with the AMPL solver library.

Solver options can be specified in a generic manner using the `--solver-options` option. This specifies a string interpreted as one or more option-value pairs. For example, the following option passes the `mipgap` option to the glpk solver:

```
pyomo solve --solver=glpk --solver-options='mipgap=0.01' \
    concrete1.py
```

Additionally, the `--timelimit` option can be used to specify the maximum runtime of the solver. This is typically passed to the solver, and thus this timelimit is enforced in a solver-dependent manner.

Solver results are generated from solution information provided by the solver, and optionally a logfile of output from the solver. By default, Pyomo captures information about the variable values selected by the solver. However, there is often additional information a user may wish to collect, such as dual values for constraints in a linear program. For performance reasons, this data is not automatically collected by the `solve` subcommand, but the `--solver-suffixes` option is used to specify the names of the data desired. A *suffix* is simply data for a constraint or variable that results from the application of a solver. Suffixes can be specified by name, or with a regular expression. For example, the following command specifies that all suffixes generated by the solver are requested:

```
pyomo solve --solver=glpk --solver-suffix='.*' concrete2.py
```

The following suffixes are currently supported within Pyomo:

- `dual` - constraint dual values
- `rc` - reduced costs
- `slack` - constraint slack values

Note that a given solver may provide only a subset of these suffixes.

The `--tempdir` and `--keepfiles` options can be used to archive the temporary files that Pyomo uses. By default, Pyomo uses temporary files automatically generated in system temporary directories. The `--tempdir` option is used to specify the directory that these files are created in. By default, temporary files are deleted after optimization is completed. The `--keepfiles` options disables this deletion, which allows the user to see the data Pyomo sends to the optimizer.

10.2.2.5 Analyze Solver Results

The `--postprocess` option can be used to specify a Python module that is executed after the solver has executed. A typical use of this option is to specify post-processing steps to interpret the solver results in a problem-dependent manner.

Post-processing steps can be defined by declaring in the Python modules a `pyomo_postprocess` function to be used in post processing. Figure 10.2 provides an example of a post-processing function that writes the final solutions to a file in the CSV format.

10.2.2.6 Managing Diagnostic Output

The `solve` subcommand includes a variety of options to control the generation of diagnostic output and other information useful for learning more about the executed workflow.

```
import csv

def pyomo_postprocess(options=None, instance=None,
                                  results=None):
    #
    # Collect the data
    #
    vars = set()
    data = {}
    f = {}
    for i in range(len(results.solution)):
        data[i] = {}
        for var in results.solution[i].variable:
            vars.add(var)
            data[i][var] = \
                results.solution[i].variable[var]['Value']
        for obj in results.solution[i].objective:
            f[i] = results.solution[i].objective[obj]['Value']
            break
    #
    # Write a CSV file, with one row per solution.
    # The first column is the function value, the remaining
    # columns are the values of nonzero variables.
    #
    rows = []
    vars = list(vars)
    vars.sort()
    rows.append(['obj']+vars)
    for i in range(len(results.solution)):
        row = [f[i]]
        for var in vars:
            row.append( data[i].get(var,None) )
        rows.append(row)
    print("Creating results file results.csv")
    OUTPUT = open('results.csv', 'w')
    writer = csv.writer(OUTPUT)
    writer.writerows(rows)
    OUTPUT.close()
```

Fig. 10.2: A post-processing plugin that writes final solutions in a CSV file.

The default output of the solve subcommand is a terse summary of the major steps that are executed. The --log and --stream-output options are used to print the solver output. The --log option is used to print the solver output after the solver has terminated, and the --stream-output option is used to print the solver output as it is generated. Similarly, the --summary and --show-results options print different summaries of the optimization results. The --summary command prints a summary of the Pyomo model, after the results are loaded.

The --show-results prints the final results. If the PyYAML package is installed, then the default results format is YAML and the final results are stored in the file `results.yml`. Otherwise, the default results format is JSON and the final results are stored in the file `results.json`. The `--json` option can be used to specify the JSON results format when the `PyYAML` package is installed. The `--save-results` option can be used to specify an alternative results file.

Pyomo uses a standard Python logging system to manage the printing of logging messages for the underlying software in Pyomo. By default, logging messages representing Pyomo errors and warnings are always printed. The `--quiet` option suppresses all log messages except for those referring to errors. The `--warning` option enables warning messages for Pyomo. The `--info` option enables informative, warning and error log messages for Pyomo.

The `--verbose` option enables debugging log messages for Pyomo. This option can be specified multiple times to enable logging messages for different parts of Pyomo: (1) debugging for just Pyomo and (2) debugging for all Pyomo packages. The `--debug` option enables debugging logging, and it allows exceptions to trigger a failure in which the program stack is printed.

10.2.3 The `convert` Subcommand

Many optimizers supported by Pyomo read a a temporary file Pyomo generates in a standard problem format. For example, the NL format is recognized by solvers used with the AMPL modeling tool, and the LP file format is used by a variety of commercial and open source integer programming solvers.

It is often useful to generate these problem files directly, both to diagnose issues with a model as well as to directly manage the execution of a solver. The `convert` subcommand can be used to convert a Pyomo model into a standard file format. For example, consider the command:

```
pyomo convert --format=lp concrete1.py
```

This command converts the model in `concrete1.py` into an LP file format, which is stored in the file `unknown.lp`:

```
[    0.00] Setting up Pyomo environment
[    0.00] Applying Pyomo preprocessing actions
[    0.00] Creating model
Model written to file 'unknown.lp'
[    0.05] Pyomo Finished
```

The `--output` option can also be used to specify a filename, for which the filename suffix specifies the file format. For example, the command

```
pyomo convert --output=concrete1.lp concrete1.py
```

creates the file `concrete1.lp`, which represents the model from `concrete1.py` in the LP file format.

The command

```
pyomo help -w
```

summarizes the file formats supported by Pyomo.

10.3 Data Commands for AbstractModel

The Set and Param components of a Pyomo model are used to define data values
used to construct constraints and objectives. Previous chapters have illustrated that
these components are not necessary to develop complex models. However, the Set
and Param components can be used to define abstract data declarations, where no
data values are specified. For example:

```
model.A = Set(within=Reals)
model.p = Param(model.A, within=Integers)
```

Data command files can be used to initialize data declarations in Pyomo models,
and in particular they are useful for initializing AbstractModel data declarations. However, note that complex mappings are often accomplished in Pyomo via
scripting rather than using data command files.

Pyomo's data command files employ a domain-specific language whose syntax
closely resembles the syntax of AMPL's data commands [2]. A data command file
consists of a sequence of commands specifing set and parameter data, or specifing
where such data is to be obtained from external sources. The following commands
can be used to declare data:

- The set command declares set data.
- The param command declares a table of parameter data, which can also include
  the declaration of the set data used to index the parameter data.
- The load command loads data from an external resources, like a spreadsheet
  or database.
- The table command declares a two-dimensional table of parameter data.

The following commands can also be used in data command files:

- The include command specifies a data command file that is processed immediately.
- The data and end commands do not perform any actions, but they provide
  compatibility with AMPL scripts that define data commands.

Finally, the namespace declaration allows data commands to be organized into
named groups allowing each to be enabled or disabled during model construction.

Note that Pyomo's data commands do *not* exactly correspond to AMPL data
commands. The set and param commands are designed to closely match AMPL's
syntax and semantics. However, these commands only support a subset of the corresponding declarations in AMPL. However, it is not possible to support other AMPL

commands because Pyomo treats data commands as data declarations while AMPL treats data commands as part of its scripting language.

The following subsections describe the syntax Pyomo's data file commands except for the `load` and `table` commands, which are documented via Pyomo's online documentation. The syntax of data commands can be quite varied, and we provide detailed examples to illustrate these commands. Note that all Pyomo data commands are terminated with a semicolon, and the syntax of data commands does not depend on whitespace. Thus, data commands can be broken across multiple lines – newlines and tab characters are ignored – and data commands can be formatted with whitespace with few restrictions.

10.3.1 The `set` Command

10.3.1.1 Simple Sets

The `set` data command explicitly specifies the members of either a single set or an array of sets, i.e., an indexed set. A single set is specified with a list of data values that are included in this set. The formal syntax for the set data command is:

```
set <setname> := [<value>] ... ;
```

The data values in a set consist of either numeric values, simple strings or quoted strings:

- *Numeric values* are any string that can be evaluated by Python as a numeric value, e.g., integer, float, scientific notation, or boolean.
- *Simple strings* are sequences of alpha-numeric characters.
- *Quoted strings* are simple strings that are included in a pair of single or double quotes. A quoted string can include quotes within the quoted string.

There is no restriction on the values in a set declaration. A set may be empty, and it may contain any combination of numeric and non-numeric string values. Validation of set data is performed when constructing a Pyomo model, not while parsing a data command file. For example, the following are valid `set` commands:

```
# An empty set
set A := ;

# A set of numbers
set A := 1 2 3;

# A set of strings
set B := north south east west;

# A set of mixed types
set C :=
0
-1.0e+10
```

```
'foo bar'
infinity
"100"
;
```

Note that numeric values are automatically converted to Python integer or floating point values when the set data specification is parsed. A quoted string can be used to define a string value containing a numeric value. However, if the string strictly specifies a numeric value, it will be converted by Python to a numeric type. For example, the string "100" is included in set C, but this value is converted to a numeric value.

10.3.1.2 Sets of Tuple Data

The set data command can also specify tuple data with the standard notation for tuples. For example, suppose set A contains 3-tuples:

```
model.A = pyo.Set(dimen=3)
```

The following set data command then specifies that A is the set containing the tuples (1,2,3) and (4,5,6):

```
set A := (1,2,3) (4,5,6) ;
```

Alternatively, set data can simply be listed in the order that the tuple is represented:

```
set A := 1 2 3 4 5 6 ;
```

Obviously, the number of data elements specified using this syntax should be a multiple of the set dimension.

Sets with 2-tuple data can also be specified in a matrix denoting set membership. For example, the following set data command declares 2-tuples in A using + to denote valid tuples and − to denote invalid tuples:

```
set A : A1 A2 A3 A4 :=
    1    +   -   -   +
    2    +   -   +   -
    3    -   +   -   - ;
```

This data command declares the following five 2-tuples: ('A1',1), ('A1',2), ('A2',3), ('A3',2), ('A4',1).

Finally, a set of tuple data can be concisely represented with tuple *templates* that represent a *slice* of tuple data. For example, suppose the set A contains 4-tuples:

```
model.A = pyo.Set(dimen=4)
```

The following set data command declares groups of tuples defined by a template and data to complete this template:

```
set A :=
    (1,2,*,4) A B
    (*,2,*,4) A B C D ;
```

A tuple template consists of a tuple containing one or more * symbols instead of a value. These represent indices where the tuple value is replaced by the values from the list of values following the tuple template. In this example, the following tuples are in set A:

```
(1,  2,  'A',  4)
(1,  2,  'B',  4)
('A',  2,  'B',  4)
('C',  2,  'D',  4)
```

10.3.1.3 Set Arrays

The `set` data command can also be used to declare data for a set array. Each set in a set array must be declared with a separate `set` data command with the following syntax:

```
set <set-name>[<index>] := [<value>] ... ;
```

Set arrays can be indexed by an arbitrary set and the index value may be a numeric value, a non-numeric string value, or a comma-separated list of string values.

Suppose set A is used to index a set B as follows:

```
model.A = pyo.Set()
model.B = pyo.Set(model.A)
```

Then set B is indexed using the values declared for set A:

```
set A := 1 aaa 'a b';

set B[1] := 0 1 2;
set B[aaa] := aa bb cc;
set B['a b'] := 'aa bb cc';
```

10.3.2 The `param` Command

Simple or non-indexed parameters are declared in an obvious way, as shown by these examples:

```
param A := 1.4;
param B := 1;
param C := abc;
param D := true;
param E := 1.0e+04;
```

Parameters can be defined with numeric and string data. Numeric data is defined with a string evaluated by Python as a numeric value, which includes integer, floating point, scientific notation, and boolean. Boolean values can be specified with a variety of strings: TRUE, true, True, FALSE, false, and False. Note that parameters cannot be defined without data, so there is no analog to the specification of an empty set.

Most parameter data is indexed over one or more sets, and there are a number of ways the `param` data command can be used to specify indexed parameter data.

10.3.2.1 One-dimensional Parameter Data

One-dimensional parameter data is indexed over a single set. Suppose parameter B is indexed by the set A:

```
model.A = pyo.Set()
model.B = pyo.Param(model.A)
```

A `param` data command can specify values for B with a list of index-value pairs:

```
set A := a c e;

param B := a 10 c 30 e 50;
```

Because whitespace is ignored, this example data command file can be reorganized to specify the same data in a tabular format:

```
set A := a c e;

param B :=
a 10
c 30
e 50
;
```

Multiple parameters can be defined using a single `param` data command. For example, suppose parameters B, C, and D are one-dimensional parameters all indexed by the set A:

```
model.A = pyo.Set()
model.B = pyo.Param(model.A)
model.C = pyo.Param(model.A)
model.D = pyo.Param(model.A)
```

Values for these parameters can be specified using a single `param` data command declaring these parameter names followed by a list of index and parameter values:

```
set A := a c e;

param : B C D :=
a 10 -1 1.1
c 30 -3 3.3
e 50 -5 5.5
;
```

The values in the `param` data command are interpreted as a list of sublists, where each sublist consists of an index followed by the corresponding numeric value.

Note that parameter values do not need to be defined for all indices. For example, the following data command file is valid:

```
set A := a c e g;

param : B C D :=
a 10 -1 1.1
c 30 -3 3.3
e 50 -5 5.5
;
```

The index g is omitted from the `param` command, and consequently this index is not valid for the model instance using this data. More complex patterns of missing data can be specified using the "`.`" character to indicate a missing value. This syntax is useful when specifying multiple parameters that do not necessarily have the same index values:

```
set A := a c e;

param : B C D :=
a  .  -1 1.1
c 30  .  3.3
e 50 -5   .
;
```

This example provides a concise representation of parameters that share a common index set while using different index values.

Note that this data file specifies the data for set A twice: (1) when A is defined and (2) implicitly when the parameters are defined. An alternate syntax for `param` allows the user to concisely specify the definition of an index set along with associated parameters:

```
param : A : B C D :=
a 10 -1 1.1
c 30 -3 3.3
e 50 -5 5.5
;
```

Finally, we note that default values for missing data can also be specified using the `default` keyword:

```
set A := a c e;

param B default 0.0 :=
c 30
e 50
;
```

Note that default values can only be specified in `param` commands defining values for a single parameter.

10.3.2.2 Multi-Dimensional Parameter Data

Multi-dimensional parameter data is indexed over either multiple sets or multi-dimensional sets. Suppose parameter B is indexed by set A with dimension 2:

```
model.A = pyo.Set(dimen=2)
model.B = pyo.Param(model.A)
```

The syntax of the `param` data command remains essentially the same when specifying values for B with a list of index and parameter values:

```
set A := a 1 c 2 e 3;

param B :=
a 1 10
c 2 30
e 3 50;
```

Missing and default values are also handled in the same way with multi-dimensional index sets:

```
set A := a 1 c 2 e 3;

param B default 0 :=
a 1 10
c 2 .
e 3 50;
```

Similarly, multiple parameters can defined with a single `param` data command. Suppose that parameters B, C, and D are parameters indexed over set A with dimension 2:

```
model.A = pyo.Set(dimen=2)
model.B = pyo.Param(model.A)
model.C = pyo.Param(model.A)
model.D = pyo.Param(model.A)
```

These parameters can be defined with a single `param` command that declares the parameter names followed by a list of index and parameter values:

```
set A := a 1 c 2 e 3;

param : B C D :=
a 1 10 -1 1.1
c 2 30 -3 3.3
e 3 50 -5 5.5
;
```

Similarly, the following `param` data command defines the index set along with the parameters:

```
param : A : B C D :=
a 1 10 -1 1.1
c 2 30 -3 3.3
e 3 50 -5 5.5
;
```

The `param` command also supports a matrix syntax for specifying the values in a parameter with a 2-dimensional index. Suppose parameter B is indexed over set A with dimension 2:

```
model.A = pyo.Set(dimen=2)
model.B = pyo.Param(model.A)
```

The following `param` command defines a matrix of parameter values:

```
set A := 1 a 1 c 1 e 2 a 2 c 2 e 3 a 3 c 3 e;

param B : a c e :=
1 1 2 3
2 4 5 6
3 7 8 9
;
```

Additionally, the following syntax can be used to specify a transposed matrix of parameter values:

```
set A := 1 a 1 c 1 e 2 a 2 c 2 e 3 a 3 c 3 e;

param B (tr) : 1 2 3 :=
a 1 4 7
c 2 5 8
e 3 6 9
;
```

This functionality facilitates the presentation of parameter data in a natural format. In particular, the transpose syntax may allow the specification of tables for which the rows comfortably fit within a single line. However, a matrix may be divided column-wise into shorter rows since the line breaks are not significant in Pyomo's data commands.

For parameters with three or more indices, the parameter data values must be specified as a series of slices. Each slice is defined by a template followed by a list of index and parameter values. Suppose that parameter B is indexed over set A that has dimension 4:

```
model.A = pyo.Set(dimen=4)
model.B = pyo.Param(model.A)
```

The following `param` command defines a matrix of parameter values with multiple templates:

```
set A := (a,1,a,1) (a,2,a,2) (b,1,b,1) (b,2,b,2);

param B :=

  [*,1,*,1] a a 10 b b 20
  [*,2,*,2] a a 30 b b 40
;
```

The B parameter consists of four values: B[a,1,a,1]=10, B[b,1,b,1]=20, B[a,2,a,2]=30, and B[b,2,b,2]=40.

10.3.3 The `include` Command

The `include` command allows a data command file to execute data commands from another file. For example, the following command file executes data commands from `ex1.dat` and then `ex2.dat`:

```
include ex1.dat;
include ex2.dat;
```

Pyomo is sensitive to the order of execution of data commands, since data commands can redefine set and parameter values. The `include` command respects this data ordering; all data commands in the included file are executed before the remaining data commands in the current file are executed.

10.3.4 Data Namespaces

The `namespace` keyword is not a data command, but instead it is used to structure the specification of Pyomo's data commands. Specifically, a namespace declaration is used to group data commands and to provide a group label. Consider the following data command file:

```
set C := 1 2 3 ;

namespace ns1
{
    set C := 4 5 6 ;
}

namespace ns2
{
    set C := 7 8 9 ;
}
```

This data file defines two namespaces: `ns1` and `ns2` that initialize a set C. By default, data commands contained within a namespace are ignored during model construction; when no namespaces are specified, the set C has values 1, 2, 3. When namespace `ns1` is specified, then set C values are overridden with the set 4, 5, 6. See Section 10.2.2.2 for an example of how namespaces are selected with the `pyomo` command.

10.4 Build Components

In a function constructing a `ConcreteModel` one can insert Python code anywhere in the process. One can, for example, fix a particular combination of variables, print the value of a parameter, or throw an exception if a particular combination of parameter values is not valid. To provide this functionality for an `AbstractModel`, Pyomo supports a set of components enabling execution of Python code during the build process.

The `BuildAction` component can be defined in the model to inject actions (defined through Python code) into the model construction process. Similarly, the `BuildCheck` component is used to perform a user-defined test (again, through Python code) during the model construction process and halt construction if the test fails. These components are added to a model in the same manner as other components, but their role is to allow a user to insert scripting-like code into the model construction process.

Consider the following abstract model (defined in `buildactions.py`) illustrating the use of `BuildAction` and `BuildCheck` components to define error checks and diagnostic output based on our warehouse location example defined in Sections 3.2 and 10.1.3.

```python
# buildactions.py: Warehouse location problem showing build \
    actions
import pyomo.environ as pyo

model = pyo.AbstractModel()

model.N = pyo.Set() # Set of warehouses
model.M = pyo.Set() # Set of customers
model.d = pyo.Param(model.N,model.M)
model.P = pyo.Param()

model.x = pyo.Var(model.N, model.M, bounds=(0,1))
model.y = pyo.Var(model.N, within=pyo.Binary)

def checkPN_rule(model):
    return model.P <= len(model.N)
model.checkPN = pyo.BuildCheck(rule=checkPN_rule)

def obj_rule(model):
    return sum(model.d[n,m]*model.x[n,m] for n in model.N for m \
        in model.M)
model.obj = pyo.Objective(rule=obj_rule)

def one_per_cust_rule(model, m):
    return sum(model.x[n,m] for n in model.N) == 1
model.one_per_cust = pyo.Constraint(model.M, \
    rule=one_per_cust_rule)

def warehouse_active_rule(model, n, m):
    return model.x[n,m] <= model.y[n]
model.warehouse_active = pyo.Constraint(model.N, model.M, \
    rule=warehouse_active_rule)

def num_warehouses_rule(model):
    return sum(model.y[n] for n in model.N) <= model.P
model.num_warehouses = pyo.Constraint(rule=num_warehouses_rule)
```

```
def printM_rule(model):
   model.M.pprint()
model.printM = pyo.BuildAction(rule=printM_rule)
```

In this example, we have added a `BuildCheck` component with the rule `CheckPN_rule`. This rule will check to make sure the total number of warehouses we can place is not more than the number of available warehouse locations. We have also added a `BuildAction` component with the rule `printM_rule` to print the set of customer locations.

We created a .dat file where the parameter *P* is larger than the available number of warehouse locations (so it would fail the `CheckPN_rule` build check:

```
# buildactions_fails.dat: Pyomo format data file for the \
    warehouse location problem
# Note: parameter P is larger than the number of warehouse \
    locations

set N := Harlingen Memphis Ashland ;
set M := NYC LA Chicago Houston;

param d :=
    Harlingen NYC 1956
    Harlingen LA   1606
    Harlingen Chicago   1410
    Harlingen Houston   330
    Memphis NYC   1096
    Memphis LA   1792
    Memphis Chicago   531
    Memphis Houston   567
    Ashland NYC   485
    Ashland LA   2322
    Ashland Chicago   324
    Ashland Houston   1236
;

param P := 4 ;
```

Solving this with the `pyomo` command:

```
pyomo solve --solver=glpk buildactions.py buildactions_fails.dat
```

gives us output similar to the following:

```
[     0.00] Setting up Pyomo environment
[     0.00] Applying Pyomo preprocessing actions
[     0.00] Creating model
ERROR: Constructing component 'checkPN' from data=None failed:
        ValueError: BuildCheck 'checkPN' identified error
[     0.01] Pyomo Finished
ERROR: Unexpected exception while running model:
        BuildCheck 'checkPN' identified error
```

As with other components, the `BuildAction` and `BuildCheck` components can be indexed, which allows actions and checks to be customized based on specific data.

Part III
Modeling Extensions

Chapter 11
Generalized Disjunctive Programming

Abstract This chapter documents how to express and solve Generalized Disjunctive Programs (GDPs). GDP models provide a structured approach for describing logical relationships in optimization models. We show how Pyomo blocks provide a natural base for representing disjuncts and forming disjunctions, and we how to solve GDP models through the use of automated problem transformations.

11.1 Introduction

A common feature of many discrete optimization problems is the selection among two or more discrete choices. These decisions imply additional restrictions on the feasible problem space. For example, a *semi-continuous variable* either has value zero or must be above a given value. This can be expressed algebraically as either $x = 0$ or $L \leq x \leq U$, where $L > 0$ and U is allowed to be infinite. When U is finite, this property can be enforced by defining a binary variable y and defining the following constraints:

$$Ly \leq x$$
$$x \leq Uy$$
$$y \in \{0,1\}$$

When U is infinite, then the second inequality is replaced with a so-called "Big-M" constraint:

$$x \leq My$$

for a value M chosen to be sufficiently big so as to not limit the space of feasible solutions (although it does imply a finite U). One can readily see that when $y = 1$ the inequalities enforce that x lie in the continuous range $L \leq x \leq U$, and when $y = 0$ the inequalities reduce to $0 \leq x \leq 0$, or $x = 0$.

In this formulation, the binary variable y is not intrinsic to the constraints, but rather captures a logical condition: indicating whether one set of mututally exclusive constraints ($L \leq x \leq U$) or another ($x = 0$) must be enforced. In this way, the variable is often referred to as an *indicator variable*. Indeed, it can be argued that the bulk of binary variables used in math programming are in fact indicator variables used to capture logic in the model.

This approach to formulating a switching decision can be generalized to switch between different groups of constraints. For example, consider the Unit Commitment Problem, where the goal of the optimization problem is to determine the minimal cost schedule for turning on and off a fleet of generators in order to meet expected demand. In this case, a generator can exist in one of several states: on, off, starting up, and shutting down. Selecting a particular state implies additional constraints on the generator operation. If a generator is "on", then the output power is bounded between the minimum (nonzero) and maximum power levels. In contrast, when the generator is "off", the output power must be 0. Further, the output power in any time period must be within "ramp limits" of output power in the previous time period.

One implementation of these state selection rules expresses all the constraints and relaxes them based on binary state (indicator) variables using so-called "Big-M" terms:

$$Power_{g,t} \leq MaxPower_g \cdot GenOn_{g,t} \quad (11.1)$$

$$Power_{g,t} \geq MinPower_g \cdot GenOn_{g,t} \quad (11.2)$$

$$Power_{g,t} \leq Power_{g,t-1} + RampUpLimit_g + M_g \cdot (1 - GenOn_{g,t}) \quad (11.3)$$

$$Power_{g,t} \geq Power_{g,t-1} - RampDownLimit_g - M_g \cdot (1 - GenOn_{g,t}) \quad (11.4)$$

$$Power_{g,t} \leq MaxPower_g \cdot (1 - GenOff_{g,t}) \quad (11.5)$$

$$Power_{g,t-1} \leq ShutDownRampLimit_g + MaxPower_g \cdot (1 - GenOff_{g,t}) \quad (11.6)$$

$$Power_{g,t} \leq StartUpRampLimit_g + MaxPower_g \cdot (1 - GenStartUp_{g,t}) \quad (11.7)$$

$$GenOn_{g,t} + GenOff_{g,t} + GenStartUp_{g,t} = 1 \quad (11.8)$$

$$GenOn_{g,t} \leq GenOn_{g,t-1} + GenStartUp_{g,t-1} \quad (11.9)$$

$$GenStartUp_{g,t} \leq GenOff_{g,t-1} \quad (11.10)$$

$$GenOn_{g,t}, GenOff_{g,t}, GenStartUp_{g,t}, GenShutDown_{g,t} \in \{0,1\} \quad (11.11)$$

$$Power_{g,t} \in (0, MaxPower_g) \quad (11.12)$$

This approach to formulating a switching decision has two significant limitations. First, the relationships between the binary selection variable and the corresponding constraints that the binary variable selects is somewhat obfuscated. Second, the use of "Big-M" relaxations is only one of several possible approaches to formulating the problem. By hard-coding that relaxation into your model, you are effectively precluding the possibility of exploring alternative approaches (like a hull relaxation) without incurring significant effort rewriting the model.

Generalized Disjunctive Programming [54] represents an alternative approach to representing problems with significant logical structure. It generalizes the concepts of Disjunctive Programming [6] for integer linear problems to also include nonlinear systems. The canonical GDP model [38] augments the objective, variables, and constraints of a typical MI(N)LP with Boolean variables, disjunctions, and logical constraints:

$$\min \quad \sum_{k \in K} c_k + f(x) \tag{11.13}$$

$$s.t. \quad r(x) \leq 0 \tag{11.14}$$

$$\bigvee_{j \in J_k} \begin{bmatrix} Y_{jk} \\ g_{jk}(x) \leq 0 \\ c_k = \gamma_{j,k} \end{bmatrix} \quad \forall k \in K \tag{11.15}$$

$$\Omega(Y) = True \tag{11.16}$$

$$x \geq 0, c_k \geq 0, Y_{jk} \in \{True, False\} \tag{11.17}$$

In this framework, the logical decisions are represented as sets of disjunctions (Eqn. 11.15) and logical constraints (Eqn. 11.16). Each disjunction contains a number of terms (*disjuncts*) connected by an "OR" operator. Each disjunct contains a Boolean indicator variable (Y) and a set of constraints only enforced when Y is *True*. Additional constraints enforcing logical relationships among the indicator variables are imposed through Eqn. 11.16.

Recasting the generator state model as a GDP yields the following disjunction:

$$\begin{bmatrix} Y_{g,on} \\ MinPower_g \leq Power_{g,t} \leq MaxPower_g \\ -RampDownLimit_g \leq Power_{g,t} - Power_{g,t-1} \leq RampUpLimit_g \end{bmatrix}$$

$$\vee \begin{bmatrix} Y_{g,off} \\ Power_{g,t} = 0 \\ Power_{g,t-1} \leq ShutDownRampLimit_g \end{bmatrix}$$

$$\vee \begin{bmatrix} Y_{g,startup} \\ Power_{g,t} \leq StartUpRampLimit_g \end{bmatrix} \tag{11.18}$$

This modeling approach directly addresses the two limitations of typical MI(N)LP models discussed previously: the relationship between the switching variable and the constraints it implies is now explicit in the model structure, and the model is no longer locked into any particular relaxation.

11.2 Modeling GDP in Pyomo

The `pyomo.gdp` package extends the core modeling environment to represent GDP models. This package defines two new constructs: the *disjunct* and the *disjunction*. We implement these constructs as two new components in `pyomo.gdp`: `Disjunct` and `Disjunction`, respectively. The components are imported from the GDP package:

```
from pyomo.gdp import Disjunct, Disjunction
```

A disjunct is logically a container for the indicator variable and the corresponding constraints. Here we see the power of the hierarchical modeling approach enabled by the `Block` component: the `Disjunct` component is naturally derived from the `Block` class. As with blocks, `Disjunct` components may be arbitrarily indexed and initialized through rules. In addition, they may contain any Pyomo modeling component, including not only `Sets`, `Params`, `Vars`, and `Constraints`, but also `Blocks`, `Disjuncts`, and `Disjunctions`. The only thing that the `Disjunct` class adds to the normal `Block` implementation is the implicit and automatic definition of the disjunct's Boolean indicator variable.

For our generator state example, the requisite three disjuncts are declared as follows:

```
model.NumTimePeriods = pyo.Param()
model.GENERATORS = pyo.Set()
model.TIME = pyo.RangeSet(model.NumTimePeriods)

model.MaxPower = pyo.Param(model.GENERATORS, \
    within=pyo.NonNegativeReals)
model.MinPower = pyo.Param(model.GENERATORS, \
    within=pyo.NonNegativeReals)
model.RampUpLimit = pyo.Param(model.GENERATORS, \
    within=pyo.NonNegativeReals)
model.RampDownLimit = pyo.Param(model.GENERATORS, \
    within=pyo.NonNegativeReals)
model.StartUpRampLimit = pyo.Param(model.GENERATORS, \
    within=pyo.NonNegativeReals)
model.ShutDownRampLimit = pyo.Param(model.GENERATORS, \
    within=pyo.NonNegativeReals)

def Power_bound(m,g,t):
    return (0, m.MaxPower[g])
model.Power = pyo.Var(model.GENERATORS, model.TIME, \
    bounds=Power_bound)

def GenOn(b, g, t):
    m = b.model()
    b.power_limit = pyo.Constraint(
        expr=pyo.inequality(m.MinPower[g], m.Power[g,t], \
            m.MaxPower[g]) )
    if t == m.TIME.first():
        return
    b.ramp_limit = pyo.Constraint(
```

```
          expr=pyo.inequality(-m.RampDownLimit[g],
                          m.Power[g,t] - m.Power[g,t-1],
                          m.RampUpLimit[g])
    )
model.GenOn = Disjunct(model.GENERATORS, model.TIME, rule=GenOn)

def GenOff(b, g, t):
    m = b.model()
    b.power_limit = pyo.Constraint(
        expr=m.Power[g,t] == 0 )
    if t == m.TIME.first():
        return
    b.ramp_limit = pyo.Constraint(
        expr=m.Power[g,t-1] <= m.ShutDownRampLimit[g] )
model.GenOff = Disjunct(model.GENERATORS, model.TIME, \
    rule=GenOff)

def GenStartUp(b, g, t):
    m = b.model()
    b.power_limit = pyo.Constraint(
        expr=m.Power[g,t] <= m.StartUpRampLimit[g] )
model.GenStartup = Disjunct(model.GENERATORS, model.TIME, \
    rule=GenStartUp)
```

Note that while the disjuncts may be completely self-contained, with their own local variables, parameters, and constraints, they may also reference Pyomo components outside their immediate scope.

The Disjunction component is used to associate a set of disjuncts. A disjunction is similar to a constraint, in that it can be indexed and defined through rules. However, unlike a Constraint, where the rule returns a relational expression, the rule for a Disjunction must return a list of Disjuncts. Subsequent model transformations will convert the Disjunction component and generate the binding constraint across the disjuncts. While the general form of a GDP relates the disjuncts using an "OR" operator, the vast majority of models actually expect an "exactly one" relationship (a generalization of the "exclusive OR" operator). This is so common that the default behavior of the Disjunction component is to generate the "exactly one" relationship. Modelers may, however, specify the original "OR" operator by providing xor=False to the Disjunction declaration.

The disjunction for our generator state example is properly an "exactly one" relationship, and can be expressed in Pyomo using:

```
def bind_generators(m, g, t):
    return [m.GenOn[g, t], m.GenOff[g, t], m.GenStartup[g, t]]
model.bind_generators = Disjunction(
    model.GENERATORS, model.TIME, rule=bind_generators)
```

11.3 Expressing logical constraints

While `Disjunct` blocks capture the indicator relationship between the Boolean `indicator_var` and the constraints contained on a disjunct, we also wish to express additional logical constraints among the disjunct indicator variables. To support this, Pyomo provides a `LogicalConstraint` component for representing logical constraints in a model. As is the case with `Constraint` components, `LogicalConstraint` components may be single or indexed, and are initialized using an explit `expr` keyword argument, or passed a rule function through the `rule` keyword. Where they differ is in the type of expression that they take: whereas `Constraint` accepts *relational expressions* that are equality or inequality expressions, `LogicalConstraint` accepts a *logical expression*.

Logical epressions are built using an expression system with Boolean variables and Boolean constants combined using logical operators. Table 11.1 shows the operators supported for Logical Expressions. The operations for logical expressions intentionally *do not* overlap with the operations supported for algebraic expressions, thereby enforcing a semantic distinction between *binary* (algebraic) variables and *Boolean* (logical) variables.

Table 11.1: Supported operations for generating logical expressions. In these examples, X, Y, and Z are declared `BooleanVar` variables (the `model.` is omitted due to space limitations).

Operation	Function	Method	Operator
Unary operations			
negation	`pyo.lnot(X)`		~X
Binary operations			
conjunction		`X.land(Y)`	
disjunction		`X.lor(Y)`	
exclusive disjunction	`pyo.xor(X, Y)`	`X.xor(Y)`	
implication	`pyo.implies(X, Y)`	`X.implies(Y)`	
equivalence	`pyo.equivalent(X, Y)`	`X.equivalent_to(Y)`	
***n*-ary operations**			
conjunction	`pyo.land(X, Y, Z)`		
disjunction	`pyo.lor(X, Y, Z)`		
counting (at least *m*)	`pyo.atleast(m, X, Y, Z)`		
counting (at most *m*)	`pyo.atmost(m, X, Y, Z)`		
counting (exactly *m*)	`pyo.exactly(m, X, Y, Z)`		

For our generator state example from the previous section, we will use logical constraints to capture the state transition rules describing how the generator state can change between two time periods. We can express the switching rules as:

```
def onState(m, g, t):
    if t == m.TIME.first():
        return pyo.LogicalConstraint.Skip
    return m.GenOn[g, t].indicator_var.implies(pyo.lor(
        m.GenOn[g, t-1].indicator_var,
        m.GenStartup[g, t-1].indicator_var))
model.onState = pyo.LogicalConstraint(
    model.GENERATORS, model.TIME, rule=onState)

def startupState(m, g, t):
    if t == m.TIME.first():
        return pyo.LogicalConstraint.Skip
    return m.GenStartUp[g, t].indicator_var.implies(
        m.GenOff[g, t-1].indicator_var)
model.startupState = pyo.LogicalConstraint(
    model.GENERATORS, model.TIME, rule=startupState)
```

11.4 Solving GDP models

While special-purpose solvers are being developed in Pyomo able to parse and manipulate generalized disjunctive programming models, Pyomo's standard solver interfaces cannot express directly either disjunctions or logical constraints. However, Pyomo includes the capability to *transform* a disjunctive model into an equivalent MI(N)LP model by converting the logical constraints into their equilvaent linear forms and relaxing the disjunctive constraints. The transformed (relaxed) model can then be solved by an appropriate solver through the standard solver interfaces. Pyomo's GDP package provides two automated relaxations: the first relaxes the disjunctive constraints by adding so-called "Big-M" terms (recovering the original model structure from Section 11.1) and the second explicitly generates the hull relaxation of the individual disjunctions.

11.4.1 Big-M transformation

The Big-M transformation performs a constraint-by-constraint relaxation of the original disjunctive model. This preserves the size (number of variables and constraints) of the original model at the expense of possibly generating a weak continuous relaxation.

This transformation begins by recasting each disjunct as a normal block, modifying the individual constraints to add the Big-M term. For equality constraints and 2-sided inequality constraints (those with both upper and lower bounds), the transformation duplicates the constraint as two one-sided inequality constraints before relaxing each. The values of the M parameters can be specified through a `BigM` `Suffix` placed on the `Disjunct`. When transforming linear constraints over bounded variables, this value can be estimated automatically by the transformation.

Finally, the Big-M transformation recasts the `Disjunction` components as the algebraic form of the equivalent logical constraint; that is either

$$\exists_k (d_k.indicator\_var \wedge \neg \exists_l (d_l.indicator\_var \wedge k \neq l)) \tag{11.19}$$

for exclusive disjunctions (the default), or

$$\exists_k (d_k.indicator\_var) \tag{11.20}$$

for non-exclusive disjunctions. These constraints have the equivalent algebraic forms of

$$\sum_{k \in K} b_k = 1 \tag{11.21}$$

and

$$\sum_{k \in K} b_k \geq 1 \tag{11.22}$$

where b_k is the binary variable associated with the Boolean variable $d_k.indicator\_var$. The transformation name `gdp.bigm` is used to apply the Big-M transformation.

11.4.2 Hull transformation

The hull transformation relaxes the original disjunctive model by generating a lifted representation. The transformation follows the procedure of Balas [6] for linear disjunctions and Lee and Grossmann [38] (with modifications from Sawaya and Grossmann [55]) for nonlinear disjunctions. In both cases, the variables appearing in each disjunct are "disaggregated" by defining new variables for each disjunct and constraining the original variable to be the sum of the disaggregated variables. This has the effect of representing the disjunction as the affine combination of the individual disjuncts. For disjunctions with only convex constraints, the affine combination of the disaggregated (lifted) disjuncts defines the convex hull of the disjuncts. The logical Disjunction relationship is converted to algebraic form using the same approach as Big-M (Equations 11.19-11.22).

By constraining the disaggregated variables to be zero when the disjunct's indicator variable is False, the solution to the discrete relaxed problem will be the solution to the original disjunctive problem. This increases the overall size of the model (both the number of variables and constraints), but gives a tighter continuous relaxation than the Big-M transformation. However, all variables must be bounded to apply the hull disjunction, and the Disjunctions can only express exclusive relationships (i.e.,

```
xor=True)
```
The transformation name `gdp.hull` is used to apply the Hull transformation.

11.5 A mixing problem with semi-continuous variables

The following model illustrates a simple mixing problem with three semi-continuous variables (x_1, x_2, x_3) which represent quantities that are mixed to meet a volumetric constraint. In this simple example, the number of sources is minimized:

```
# scont.py
import pyomo.environ as pyo
from pyomo.gdp import Disjunct, Disjunction

L = [1,2,3]
U = [2,4,6]
index = [0,1,2]


model = pyo.ConcreteModel()
model.x = pyo.Var(index, within=pyo.Reals, bounds=(0,20))
model.x_nonzero = pyo.Var(index, bounds=(0,1))

# Each disjunction is a semi-continuous variable
# x[k] == 0 or L[k] <= x[k] <= U[k]
def d_0_rule(d, k):
    m = d.model()
    d.c = pyo.Constraint(expr=m.x[k] == 0)
model.d_0 = Disjunct(index, rule=d_0_rule)

def d_nonzero_rule(d, k):
    m = d.model()
    d.c = pyo.Constraint(expr=pyo.inequality(L[k], m.x[k], U[k]))
    d.count = pyo.Constraint(expr=m.x_nonzero[k] == 1)
model.d_nonzero = Disjunct(index, rule=d_nonzero_rule)

def D_rule(m, k):
    return [m.d_0[k], m.d_nonzero[k]]
model.D = Disjunction(index, rule=D_rule)

# Minimize the number of x variables that are nonzero
model.o = pyo.Objective(
    expr=sum(model.x_nonzero[k] for k in index))

# Satisfy a demand that is met by these variables
model.c = pyo.Constraint(
    expr=sum(model.x[k] for k in index) >= 7)
```

There are three ways to apply either the Big-M or Hull transformation to solve this model:

1. through the `pyomo` command line,
2. through a scripting interface, or

3. through a `BuildAction`.

On the pyomo command line, the `--transform` command line option is used to apply a transformation:

```
pyomo solve scont.py --transform gdp.bigm --solver=glpk
```

The equivalent approach when developing custom scripts is to create the transformation before applying it to the model:

```
xfrm = pyo.TransformationFactory('gdp.bigm')
xfrm.apply_to(model)

solver = pyo.SolverFactory('glpk')
status = solver.solve(model)
```

Finally, there are situations where you will want to inject transformations into models that are generated and manipulated in environments other than the pyomo command or custom scripts (e.g., the runph script). In this case, you can trigger the transformation by adding a `BuildAction` to the model:

```
def transform_gdp(m):
    xfrm = pyo.TransformationFactory('gdp.bigm')
    xfrm.apply_to(m)
model.transform_gdp = pyo.BuildAction(rule=transform_gdp)
```

Chapter 12
Differential Algebraic Equations

Abstract This chapter documents how to formulate and solve optimization problems with differential and algebraic equations (DAEs). The `pyomo.dae` package
allows users to incorporate detailed dynamic models within an optimization framework, and it is flexible enough to represent a wide variety of differential equations.
`pyomo.dae` also includes several automated solution techniques based on a simultaneous discretization approach to solve dynamic optimization problems.

12.1 Introduction

In order to develop a better understanding of real-world phenomena, scientists and
engineers often develop dynamic, or differential equation based, models. High fi-
delity simulation of these models is still an active research area in many fields, since
these simulation models can be difficult and computationally expensive But after a
model suitable for simulation has been developed, the next goal is often to optimize
a particular aspect of the dynamic system (e.g., model parameter estimates given
dynamic data, or control of the dynamic system to a desired set point). Consider the
small optimal control problem from [35]:

$$\min \quad x_3(t_f) \tag{12.1}$$
$$\text{s.t.} \quad \dot{x}_1 = x_2 \tag{12.2}$$
$$\dot{x}_2 = -x_2 + u \tag{12.3}$$
$$\dot{x}_3 = x_1^2 + x_2^2 + 0.005 \cdot u^2 \tag{12.4}$$
$$x_2 - 8 \cdot (t - 0.5)^2 + 0.5 \le 0 \tag{12.5}$$
$$x_1(0) = 0, x_2(0) = -1, x_3(0) = 0, t_f = 1 \tag{12.6}$$

where the objective is to minimize the value of x_3 at the final time point by finding
the optimal values for the input variable u. This problem includes three differential

equations as constraints, and and an inequality constraint restricting the profile of x_2 (also known as a path constraint).

While it is easy to write down optimization problems including dynamic models, solving them is hard. Off-the-shelf optimization solvers cannot handle differential equations directly. Therefore, optimization problems including differential equations as constraints, known as *dynamic optimization problems*, must be reformulated in order to be solved. Common solution approaches include single or multiple shooting methods or a full discretization approach. Regardless of the solution strategy, the implementation of the technique is often entwined with the particular model or problem being solved, which makes it time-consuming to apply these solution techniques to new dynamic optimization problems or experiment with different solution strategies on the same model.

The `pyomo.dae` package addresses several of these challenges. It provides users the ability to separate the dynamic optimization formulation from the solution strategy used to solve it. This is done by introducing modeling components for representing continuous domains and derivative terms directly. `pyomo.dae` also includes implementations of the simultaneous discretization solution technique, which can be applied automatically to a Pyomo model with differential equations.

This chapter provides a brief overview of how to use the `pyomo.dae` package. We refer the reader to Nicholson et al. [44] for a more detailed description and information about the design and novelty of this package. This package is still under active development and expansion. Please refer to the online Pyomo documentation for the most up-to-date documentation on new features.

12.2 Pyomo DAE Modeling Components

The `pyomo.dae` package defines two new components used to represent DAE models in Pyomo:

- `ContinuousSet` represents continuous domains over which a derivative can be taken, and
- `DerivativeVar` represents the derivative of a `Var` with respect to a given `ContinuousSet`.

The package is explicitly imported to access these modeling components:

```
import pyomo.environ as pyo
import pyomo.dae as dae
```

The `ContinuousSet` component functions similarly to the regular Pyomo `Set`. It can be used to index other Pyomo components such as `Var`, `Constraint`, or `Expression`. A `ContinuousSet` can be thought of as a bounded, continuous range of real values. It is often used to represent time or spatial domains. In order to construct a `ContinuousSet` you must supply numeric values representing the upper and lower bounds of the continuous domain being represented. For our optimal control example, the continuous domain t is declared as follows.

```
m.tf = pyo.Param(initialize=1)
m.t = dae.ContinuousSet(bounds=(0,m.tf))
```

A separate `ContinuousSet` must be declared for each continuous domain in the model. After declaration, it can be used to index other Pyomo components and to declare derivatives in conjunction with the `DerivativeVar` component. A `DerivativeVar` must be declared for each derivative appearing in the dynamic model. Furthermore, you can only take the derivative of a `Var` with respect to a `ContinuousSet` included as an indexing set of the variable. The variables and derivatives for our optimal control example can be declared using:

```
m.u = pyo.Var(m.t, initialize=0)
m.x1 = pyo.Var(m.t)
m.x2 = pyo.Var(m.t)
m.x3 = pyo.Var(m.t)

m.dx1 = dae.DerivativeVar(m.x1, wrt=m.t)
m.dx2 = dae.DerivativeVar(m.x2, wrt=m.t)
m.dx3 = dae.DerivativeVar(m.x3)
```

Notice that the positional argument supplied to a `DerivativeVar` component is the `Var` being differentiated. The indexing sets for a `DerivativeVar` are inherited from, and identical to, those of the `Var` being differentiated. If a variable is indexed by more than one `ContinuousSet` then the `wrt` or `withrespectto` keyword argument is used to specify the desired derivative. In addition, high-order derivatives can also be declared with the `DerivativeVar` component. For example, a second order derivative can be specified with:

```
m.dx1dt2 = dae.DerivativeVar(m.x1, wrt=(m.t, m.t))
```

Differential equations can be formulated using standard Pyomo constraints. For example, the differential equations for our optimal control example are implemented with:

```
def _x1dot(m, t):
    return m.dx1[t] == m.x2[t]
m.x1dotcon = pyo.Constraint(m.t, rule=_x1dot)

def _x2dot(m, t):
    return m.dx2[t] == -m.x2[t] + m.u[t]
m.x2dotcon = pyo.Constraint(m.t, rule=_x2dot)

def _x3dot(m, t):
    return m.dx3[t] == m.x1[t]**2 + m.x2[t]**2 + 0.005*m.u[t]**2
m.x3dotcon = pyo.Constraint(m.t, rule=_x3dot)
```

The `pyomo.dae` package does not impose a particular form or structure on the differential equations. The differential equations will by default be enforced at the boundaries of the continuous domain. Depending on the dynamic model, this might not be desired. You can use the `deactivate()` method to override the enforcement of a differential equation at one or more bounds of a continuous domain as shown here:

```
m.x1dotcon[m.t.first()].deactivate()
m.x2dotcon[m.t.first()].deactivate()
m.x3dotcon[m.t.first()].deactivate()
```

The last important aspect of any dynamic optimization problem is the specification of initial or boundary conditions. This can be achieved by fixing a `Var` or `DerivativeVar` at one of the bounds of a `ContinuousSet`. The bounds of a `ContinuousSet` can be accessed using the `first()` or `last()` accessor methods on the `ContinuousSet`. For example, the initial conditions for the optimal control example can be implemented as:

```
m.x1[0].fix(0)
m.x2[m.t.first()].fix(-1)
m.x3[m.t.first()].fix(0)
```

Alternatively, constraints can be used to specify more complex conditions such as cyclic boundary conditions.

The last pieces of our optimal control example, the objective function (12.1) and the path constraint (12.5) are implemented with:

```
m.obj = pyo.Objective(expr=m.x3[m.tf])

def _con(m, t):
    return m.x2[t] - 8*(t - 0.5)**2 + 0.5 <= 0
m.con = pyo.Constraint(m.t, rule=_con)
```

12.3 Solving Pyomo Models with DAEs

Having formulated a Pyomo model with differential equations, we now describe how to solve it. None of the optimization solvers interfaced with Pyomo can currently handle differential equations directly. The only solution technique currently included with `pyomo.dae` is a simultaneous discretization approach, also called direct transcription. This approach discretizes the continuous domains in the model and approximates the differential equations using algebraic equations defined at the discretization points. The result of this discretization transformation is a purely algebraic model that can be solved with a standard nonlinear programming solver.

There are two types of discretization schemes included in `pyomo.dae`: finite difference and collocation. The schemes differ in the algebraic equations used to approximate the derivatives, but they are applied using nearly identical syntax. A discretization is applied to a particular continuous domain and propagated to each derivative and constraint over that domain. After you specify the discretization scheme and the resolution of the discretization (number of discretization points), `pyomo.dae` will automatically add the necessary discretization points to the appropriate `ContinuousSet` and add additional constraints to the Pyomo model with the discretization equations. This has the effect of *transforming* the DAE model into an algebraic model.

NOTE: Unlike other Pyomo transformations, `pyomo.dae` discretization transformations cannot currently be applied from the `pyomo` command line, you must create a transformation object and apply a discretization transformation to your model in a Python script.

12.3.1 Finite Difference Transformation

Finite difference methods approximate the derivative at a particular point using a difference equation, and they are among the simplest discretization schemes to conceptually understand and implement. Many variations differ in the choice of points used to approximate the derivative. The backward difference method, also called implicit or backward Euler, is the most common variation. To illustrate the discretization equations associated with this method we first define the following derivative and differential equation (constraint):

$$\left(\frac{dx(t)}{dt}, f(x(t), u(t)) \right) = 0, \quad t \in [0, T]. \tag{12.7}$$

After applying the backward difference method to the continuous domain t, the resulting derivative and constraint pair is

$$\left. \frac{dx}{dt} \right|_{t_{k+1}} = \frac{x_{k+1} - x_k}{h}, \quad k = 0, \dots, N-1 \tag{12.8}$$

$$g \left(\left. \frac{dx}{dt} \right|_{t_{k+1}}, f(x_{k+1}, u_{k+1}) \right) = 0, \quad k = 0, \dots, N-1 \tag{12.9}$$

where $x_k = x(t_k)$, $t_k = kh$, and h is the step size between discretization points or the size of each finite element. When a finite difference transformation is applied to a Pyomo model, the discretization equations such as (12.8) are automatically generated and added to the Pyomo model as equality constraints.

The code required to apply the backward finite difference method to our optimal control example is as follows:

```
discretizer = pyo.TransformationFactory('dae.finite_difference')
discretizer.apply_to(m, nfe=20, wrt=m.t, scheme='BACKWARD')
```

The `nfe` keyword argument stands for "number of finite elements", and it specifies the number of discretization points that are used in the discretization. The `scheme` keyword specifies which finite difference method to apply. There currently are three finite difference schemes included in `pyomo.dae`: backward finite difference ('BACKWARD'), central finite difference ('CENTRAL'), and forward finite difference ('FORWARD').

12.3.2 Collocation Transformation

The second type of discretization included in `pyomo.dae` is collocation or more specifically, orthogonal collocation over finite elements. This approach works by first breaking a continuous domain into $N - 1$ segments known as finite elements. Over each of these segments, the profiles of the differential variables (variables whose derivatives appear in the model) are approximated using polynomials. The polynomials are defined using K collocation points appearing as discretization points within each finite element. Continuity is enforced at the finite element boundaries for the differential variables. To provide a formal, mathematical representation of this approach applied to the derivative and differential equation (12.7) we have:

$$\left. \frac{dx}{dt} \right|_{t_{ij}} = \frac{1}{h_i} \sum_{j=0}^{K} x_{ij} \frac{d\ell_j(\tau_k)}{d\tau}, \quad k = 1,\ldots,K, \; i = 1,\ldots,N-1 \tag{12.10}$$

$$0 = g\left(\left. \frac{dx}{dt} \right|_{t_{ij}}, f(x_{ik}, u_{ik}) \right), \quad k = 1,\ldots,K, \; i = 1,\ldots,N-1 \tag{12.11}$$

$$x_{i+1,0} = \sum_{j=0}^{K} \ell_j(1) x_{ij}, \quad i = 1,\ldots,N-1 \tag{12.12}$$

where $t_{ij} = t_{i-1} + \tau_j h_i$, $x(t_{ij}) = x_{ij}$. Further, we note that the solution $x(t)$ is interpolated as follows:

$$x(t) = \sum_{j=0}^{K} \ell_j(\tau) x_{ij}, \quad t \in [t_{i-1}, t_i], \quad \tau \in [0,1] \tag{12.13}$$

$$\ell_j(\tau) = \prod_{k=0,\neq j}^{K} \frac{(\tau - \tau_k)}{(\tau_j - \tau_k)}. \tag{12.14}$$

Collocation methods produce significantly more accurate algebraic approximations compared to finite difference methods. However, they are much harder to implement manually. Variations of collocation methods differ in the functional representation of the differential variable profile over each finite element as well as the selection of the collocation points. As of this writing, the collocation transformations in `pyomo.dae` use Lagrange polynomials to represent differential variable profiles. Two options are available for the selection of collocation points: shifted Gauss-Radau roots ('LAGRANGE-RADAU') and shifted Gauss-Legendre roots ('LAGRANGE-LEGENDRE').

A collocation discretization can be applied to a Pyomo model using:

```
discretizer = pyo.TransformationFactory('dae.collocation')
discretizer.apply_to(m,nfe=7,ncp=6,scheme='LAGRANGE-RADAU')
```

The `nfe` keyword argument specifies the number of finite elements and the `ncp` argument specifies the number of collocation points within each finite element.

12.4 Additional Features

There are several advanced features included `pyomo.dae`. In this section, we briefly mention two such features that will be of interest for users interested in PDE constrained optimization or more advanced optimal control strategies.

12.4.1 Applying Multiple Discretizations

As mentioned previously, a separate discretization transformation can be applied to each `ContinuousSet` appearing in the model. This means different finite difference or collocation schemes or a combination of the two can be applied to a single Pyomo model. For example, a Pyomo model with two `ContinuousSet` components (`m.t1` and `m.t2`), could be discretized with any of the following combinations of discretization schemes:

```
# Apply multiple finite difference schemes
discretizer = pyo.TransformationFactory('dae.finite_difference')
discretizer.apply_to(m, wrt=m.t1, nfe=10, scheme='BACKWARD')
discretizer.apply_to(m, wrt=m.t2, nfe=100, scheme='FORWARD')
```

```
# Apply multiple collocation schemes
discretizer = pyo.TransformationFactory('dae.collocation')
discretizer.apply_to(m, wrt=m.t1, nfe=4, ncp=6, \
    scheme='LAGRANGE-LEGENDRE')
discretizer.apply_to(m, wrt=m.t2, nfe=10, ncp=3, \
    scheme='LAGRANGE-RADAU')
```

```
# Apply a combination of finite difference and
# collocation schemes
discretizer1 = pyo.TransformationFactory('dae.finite_difference')
discretizer2 = pyo.TransformationFactory('dae.collocation')
discretizer1.apply_to(m, wrt=m.t1, nfe=10, scheme='BACKWARD')
discretizer2.apply_to(m, wrt=m.t2, nfe=5, ncp=3, \
    scheme='LAGRANGE-RADAU')
```

12.4.2 Restricting Control Input Profiles

One of the main design considerations for the `pyomo.dae` package was the extensibility of the package to include general implementations of common operations applied to dynamic optimization problems. One such common operation in the area of optimal control is restricting the control input to have a certain profile, typically piecewise constant or piecewise linear. Often times when a model is discretized using collocation over finite elements the control variable is restricted to be constant over each finite element. The `pyomo.dae` package includes a function for doing this after a collocation discretization has been applied to a model. It works by reducing the number of free collocation points for a particular variable. For example, to restrict our control input u to be piecewise constant in our small optimal control problem you would add the following line right after applying a discretization transformation:

```
discretizer.reduce_collocation_points(m, var=m.u, ncp=1, \
    contset=m.t)
```

The `ncp` keyword argument specifies the number of free collocation points per finite element for the variable specified by the keyword `var`. Specifying `ncp=1` restricts u to have a single free collocation point (or degree of freedom) rendering it constant over each finite element. The function works by adding constraints to the discretized model which force any extra, undesired collocation points to be interpolated from the others.

12.4.3 Plotting

After formulating, discretizing, and solving a dynamic optimization problem, `pyomo.dae` makes it easy to plot the resulting dynamic profiles. Because a `ContinuousSet` is populated with floating point values from a continuous domain, the user can directly create Python lists from it for plotting. Any variable indexed by a `ContinuousSet` will have a value for each point in the `ContinuousSet`, after the model has been solved. Therefore, creating a Python list for the variable values is just as straightforward as for a `ContinuousSet`

The Python script shown below puts everything together. Assuming the Pyomo model has been declared in a separate file, the script shows how to apply a discretization and solve the model.

```
import pyomo.environ as pyo
from pyomo.dae import *
from path_constraint import m

# Discretize model using Orthogonal Collocation
discretizer = pyo.TransformationFactory('dae.collocation')
discretizer.apply_to(m,nfe=7,ncp=6,scheme='LAGRANGE-RADAU')
discretizer.reduce_collocation_points(m, var=m.u, ncp=1, \
    contset=m.t)

solver=pyo.SolverFactory('ipopt')
solver.solve(m, tee=True)
```

Finally, the code below shows an example implementation of a plotter function using `matplotlib` for plotting. The resulting figure is also shown below.

```
def plotter(subplot, x, *y, **kwds):
    plt.subplot(subplot)
    for i,_y in enumerate(y):
        plt.plot(list(x), [value(_y[t]) for t in x], 'brgcmk'[i%6])
        if kwds.get('points', False):
            plt.plot(list(x), [value(_y[t]) for t in x], 'o')
    plt.title(kwds.get('title',''))
    plt.legend(tuple(_y.name for _y in y))
    plt.xlabel(x.name)

import matplotlib.pyplot as plt
plotter(121, m.t, m.x1, m.x2, title='Differential Variables')
plotter(122, m.t, m.u, title='Control Variable', points=True)
plt.show()
```

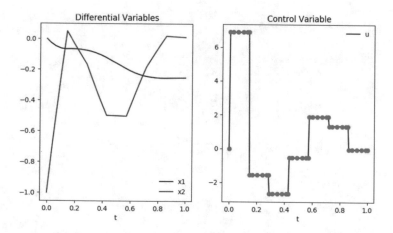

Fig. 12.1: Plot produced by `matplotlib` for the optimal control example

Chapter 13
Mathematical Programs with Equilibrium Constraints

Abstract This chapter documents how to formulate mathematical programs with equilibrium constraints (MPECs), which naturally arise in a wide range of engineering and economic systems. We describe Pyomo components for complementarity conditions, transformation capabilities that automate the reformulation of MPEC models, and meta-solvers leveraging these transformations to support global and local optimization of MPEC models.

13.1 Introduction

Mathematical Programs with Equilibrium Constraint (MPEC) problems arise in a large number of applications in engineering and economic systems [15, 40, 48]. An MPEC is an optimization problem that includes equilibrium constraints in the form of complementarity conditions. Equilibrium constraints naturally arise as the solution to an optimization subproblem (e.g., for bilevel programs), variational inequalities, and complementarity problems [29].

Since MPEC problems frequently arise in practice, many algebraic modeling languages (AML) have integrated capabilities for expressing complementarity conditions [43], including AMLs like AIMMS [1], AMPL [2, 21], GAMS [22], MATLAB [42] and YALMIP [39]. In this chapter, we describe methods for expressing and optimizing MPEC models. MPEC models can be easily expressed with Pyomo modeling components for complementarity conditions. Further, Pyomo's object-oriented design naturally supports the ability to automate the reformulation of MPEC models into other forms (e.g., disjunctive programs). We describe Pyomo meta-solvers that transform MPECs as MIP or NLP problems, which are then optimized with standard solvers. Further, we describe interfaces to specialized mixed complementarity problem solvers, which solve MPEC problems expressed without an optimization objective.

© The Author(s), under exclusive license to Springer Nature Switzerland AG 2021   191
M. L. Bynum et al., *Pyomo — Optimization Modeling in Python*, Springer Optimization
and Its Applications 67, https://doi.org/10.1007/978-3-030-68928-5_13

13.2 Modeling Equilibrium Conditions

13.2.1 Complementarity Conditions

Ferris et al. [16] note that there are a few fundamental forms accounting for a wide range of complementarity conditions that arise in practice. Consider a variable x and function $g(x)$. The classical form of complementarity condition can be expressed as

$$x \geq 0 \perp g(x) \geq 0,$$

which expresses the complementarity restriction that at least one of these must hold with equality. When the variable x is bounded such that $x \in [l, u]$, then a *mixed complementarity condition* can be expressed as

$$l \leq x \leq u \perp g(x),$$

which expresses the complementarity restriction that at least one of the following must hold:

$$
\begin{aligned}
x &= l & &\text{and } g(x) \geq 0, \\
x &= u & &\text{and } g(x) \leq 0, \\
\text{or } l < x < u & &\text{and } g(x) = 0.
\end{aligned}
$$

These forms can be generalized by substituting a function $f(x)$ for the variable x. Thus, a *generalized mixed complementarity condition* can be expressed as

$$l \leq f(x) \leq u \perp g(x),$$

which expresses the complementarity restriction that at least one of the following must hold:

$$
\begin{aligned}
f(x) &= l & &\text{and } g(x) \geq 0, \\
f(x) &= u & &\text{and } g(x) \leq 0, \\
\text{or } l < f(x) < u & &\text{and } g(x) = 0.
\end{aligned}
\tag{13.1}
$$

For completeness, note that the complementarity condition

$$f(x) \perp g(x) = 0$$

is a special case where the function $f(x)$ is unbounded.

13.2.2 Complementarity Expressions

The design of complementarity conditions in Pyomo relies on the specification of Pyomo constraint expressions. A Pyomo constraint expression defines an equality, a simple inequality, or a pair of inequalities. For example:

$$expr_1 = expr_2$$
$$expr_3 \leq expr_4$$
$$const_1 \leq expr_5 \leq const_2$$

where $const_i$ are constant arithmetic expressions that may only contain variables that are fixed, and $expr_i$ are arithmetic expressions containing unfixed variables.

A complementarity condition is defined with a pair of constraint expressions

$$l_1 \leq expr_1 \leq u_1 \perp l_2 \leq expr_2 \leq u_2,$$

where exactly two of the constant bounds l_1, u_1, l_2 and u_2 are finite. The non-finite bounds values are omitted in practice, so this condition directly describes a classical or mixed complementarity condition. Additionally, a complementarity condition can be expressed with a simple inequality, such as:

$$l_1 \leq expr_1 \perp expr_3 \leq expr_4.$$

This complementarity condition is implicitly transformed to a form with constant bounds:

$$l_1 \leq expr_1 \perp expr_3 - expr_4 \leq 0.$$

If a simple inequality is used, then the other constraint expression must also be a simple inequality to ensure the complementarity condition has exactly two finite bounds.

13.2.3 Modeling Mixed-Complementarity Conditions

Pyomo's `pyomo.mpec` package defines the `Complementarity` component that is used to declare complementarity conditions.

For example, consider the `ralph1` problem in MacMPEC [41]:

$$\min 2x - y$$
$$0 \leq y \perp y \geq x$$
$$x, y \geq 0$$

The following script defines a Pyomo model for `ralph1`:

```
# ralph1.py
import pyomo.environ as pyo
from pyomo.mpec import Complementarity, complements

model = pyo.ConcreteModel()

model.x = pyo.Var( within=pyo.NonNegativeReals )
model.y = pyo.Var( within=pyo.NonNegativeReals )

model.f1 = pyo.Objective( expr=2*model.x - model.y )

model.compl = Complementarity(
            expr=complements(0 <= model.y,
                             model.y >= model.x) )
```

The first lines in this script import Pyomo packages. The `pyomo.environ` packages initializes Pyomo's environment, and `pyomo.mpec` defines modeling components for complementarity conditions. The subsequent lines in this script create a model, declare variables x and y, declare an objective `f1`, and declare a complementarity condition `compl`.

The complementarity condition is declared with the `Complementarity` component. In the simplest case, this Python class takes a keyword argument `expr` containing the value of the `complements` function. This function accepts two Pyomo constraint expressions used to declare a complementarity condition.

Pyomo also supports indexed components, where a set of components are initialized over an index set using a construction rule. Thus, the `Complementarity` component can be declared with an index set. For example, consider the following model:

$$\min \sum_{i=1}^{n} i(x_i - 1)^2$$
$$0 \le x_i \perp 0 \le x_{i+1} \quad i = 1, \dots, n-1$$

The following script defines a Pyomo implementation of the model with $n = 5$:

```
# ex1a.py
import pyomo.environ as pyo
from pyomo.mpec import Complementarity, complements

n = 5

model = pyo.ConcreteModel()

model.x = pyo.Var( range(1,n+1) )

model.f = pyo.Objective(expr=sum(i*(model.x[i]-1)**2
                    for i in range(1,n+1)) )

def compl_(model, i):
    return complements(model.x[i] >= 0, model.x[i+1] >= 0)
model.compl = Complementarity( range(1,n), rule=compl_ )
```

The complementarity conditions are defined with a single `Complementarity`

component indexed over the set $1, \ldots, n-1$ and initialized with a construction rule
`compl_`. This rule is a function that accepts a model instance and an index, and
returns the i-th complementarity condition.

The declared set of indexes may be a superset of the indices defining complementarity conditions. If a construction rule returns `Complementarity.Skip`, then the corresponding index is skipped. For example:

```
# ex1d.py
import pyomo.environ as pyo
from pyomo.mpec import Complementarity, complements

n = 5

model = pyo.ConcreteModel()

model.x = pyo.Var( range(1,n+1) )

model.f = pyo.Objective(expr=sum(i*(model.x[i]-1)**2
                    for i in range(1,n+1)) )

def compl_(model, i):
    if i == n:
        return Complementarity.Skip
    return complements(model.x[i] >= 0, model.x[i+1] >= 0)
model.compl = Complementarity( range(1,n+1), rule=compl_ )
```

This example can also be expressed with the `ComplementarityList` component:

```
# ex1b.py
import pyomo.environ as pyo
from pyomo.mpec import ComplementarityList, complements

n = 5

model = pyo.ConcreteModel()

model.x = pyo.Var( range(1,n+1) )

model.f = pyo.Objective(expr=sum(i*(model.x[i]-1)**2
                    for i in range(1,n+1)) )

model.compl = ComplementarityList()
model.compl.add(complements(model.x[1]>=0, model.x[2]>=0))
model.compl.add(complements(model.x[2]>=0, model.x[3]>=0))
model.compl.add(complements(model.x[3]>=0, model.x[4]>=0))
model.compl.add(complements(model.x[4]>=0, model.x[5]>=0))
```

This component defines a list of complementarity conditions. The list index can be used in Pyomo, but this component simplifies the declaration of models for which the index values are not important. The `ComplementarityList` component can also be defined with a rule iteratively yielding complementarity conditions:

```
# ex1c.py
import pyomo.environ as pyo
from pyomo.mpec import ComplementarityList, complements

n = 5

model = pyo.ConcreteModel()

model.x = pyo.Var( range(1,n+1) )

model.f = pyo.Objective(expr=sum(i*(model.x[i]-1)**2
                    for i in range(1,n+1)) )

def compl_(model):
    yield complements(model.x[1] >= 0, model.x[2] >= 0)
    yield complements(model.x[2] >= 0, model.x[3] >= 0)
    yield complements(model.x[3] >= 0, model.x[4] >= 0)
    yield complements(model.x[4] >= 0, model.x[5] >= 0)
model.compl = ComplementarityList( rule=compl_ )
```

Similarly, the construction rule may be a Python list comprehension that generates a sequence of complementarity conditions:

```
# ex1e.py
import pyomo.environ as pyo
from pyomo.mpec import ComplementarityList, complements

n = 5

model = pyo.ConcreteModel()

model.x = pyo.Var( range(1,n+1) )

model.f = pyo.Objective(expr=sum(i*(model.x[i]-1)**2
                    for i in range(1,n+1)) )

model.compl = ComplementarityList(
    rule=(complements(model.x[i] >= 0, model.x[i+1] >= 0)
        for i in range(1,n)) )
```

13.3 MPEC Transformations

Pyomo supports the automated transformation of models. Pyomo can iterate through model components as well as nested model blocks. Thus, model components can be easily transformed locally, and global data can be collected to support global transformations. Further, Pyomo components and blocks can be activated and deactivated, which facilitates *in place* transformations that do not require the creation of a separate copy of the original model.

Pyomo's pyomo.mpec package defines several model transformations that can be easily applied. For example, if model defines an MPEC model (as in our previous examples), then the following example illustrates how to apply a model transformation:

```
xfrm = pyo.TransformationFactory("mpec.simple_nonlinear")
transformed = xfrm.create_using(model)
```

In this case, the mpec.simple_nonlinear transformation is applied. The following sections describe the transformations currently supported in pyomo.mpec.

13.3.1 Standard Form

In Pyomo, a complementarity condition is expressed a pair of constraint expressions

$$l_1 \leq expr_1 \leq u_1 \perp l_2 \leq expr_2 \leq u_2,$$

where exactly two of the constant bounds l_1, u_1, l_2 and u_2 are finite. The non-finite bounds are typically omitted, but the value None can be used to express infinite bounds. Additionally, each constraint expression can be expressed with a simple inequality of the form

$$expr_3 \leq expr_4.$$

The mpec.standard_form transformation reformulates each complementarity condition in a model into a standard form:

$$l_1 \leq expr \leq u_1 \perp l_2 \leq var \leq u_2,$$

where exactly two of the constant bounds l_1, u_1, l_2 and u_2 are finite, and either l_2 is zero or both l_2 or u_2 are finite.

Note that this transformation creates new variables and constraints as part of this transformation. For example, the complementarity condition

$$1 \leq x+y \perp 1 \leq 2x-y,$$

is transformed to:

$$1 \leq x+y \perp 0 \leq v,$$

where $v \in \mathbb{R}$ and $v = 2x - y - 1$.

For each complementary condition object, the new variable and constraints are added as additional components within the complementarity object. Thus, the overall structure of the MPEC model is not changed by this transformation.

13.3.2 Simple Nonlinear

The `mpec.simple_nonlinear` transformation begins by applying the `mpec.standard_form` transformation. Subsequently, a nonlinear constraint is created that defines the complementarity condition. This is a simple nonlinear transformation adapted from Ferris et al. [17], which can be described by three different cases:

- If l_1 is finite, then the following constraint is defined:

$$(expr - l_1) * v \leq \varepsilon$$

- If u_1 is finite, then the following constraint is defined:

$$(u_1 - expr) * v \leq \varepsilon$$

- If l_2 and u_2 are both finite, then the following constraints are defined:

$$(var - l_2) * expr \leq \varepsilon$$
$$(var - u_2) * expr \leq \varepsilon$$

Each of these cases ensure the complementarity condition is met when ε is zero. For example, in the first case, we know $0 \leq v$ and $0 \leq expr - l_1$. When ε is zero, this constraint ensures either v is zero or $expr - l_1$ is zero.

This transformation uses the parameter `mpec_bound`, which defines the value for ε for every complementarity condition. This allows for the specification of a relaxed nonlinear problem, which may be easier to optimize with some nonlinear programming solvers. The default value of `mpec_bound` is zero.

13.3.3 Simple Disjunction

The `mpec.simple_disjunction` transformation expresses a complementarity condition as a disjunctive program. We are given a complementarity condition defined with a pair of constraint expressions

$$l_1 \leq expr_1 \leq u_1 \perp l_2 \leq expr_2 \leq u_2,$$

where exactly two of the constant bounds l_1, u_1, l_2 and u_2 are finite. Without loss of generality, we assume that either l_1 or u_1 is finite.

This transformation generates the constraints corresponding to the conditions implied by the complementarity conditions (see Equation (13.1)). There are three different cases:

- If the first constraint is an equality, then the complementarity condition is trivially replaced by that equality constraint.

- If both bounds on the first constraint are finite but different, then the disjunction has the form:

$$\begin{bmatrix} Y_1 \\ l_1 = expr_1 \\ expr_2 \geq 0 \end{bmatrix} \bigvee \begin{bmatrix} Y_2 \\ expr_1 = u_1 \\ expr_2 \leq 0 \end{bmatrix} \bigvee \begin{bmatrix} Y_3 \\ l_1 \leq expr_1 \leq u_1 \\ expr_2 = 0 \end{bmatrix}$$

$$Y_1 \veebar Y_2 \veebar Y_3 = True$$

$$Y_1, Y_2, Y_3 \in \{True, False\}$$

- Otherwise, each constraint is a simple inequality. The complementarity condition is reformulated as

$$0 \leq \overline{expr}_1 \perp 0 \leq \overline{expr}_2,$$

and the disjunction has the form:

$$\begin{bmatrix} Y \\ 0 = \overline{expr}_1 \\ 0 \leq \overline{expr}_2 \end{bmatrix} \bigvee \begin{bmatrix} \neg Y \\ 0 \leq \overline{expr}_1 \\ 0 = \overline{expr}_2 \end{bmatrix}$$

$$Y \in \{True, False\}$$

This transformation makes use of modeling components and transformations from Pyomo's `pyomo.gdp` package. The transformation expresses each of the disjunctive terms explicitly using `Disjunct` components and the *select exactly one* logical condition using the `Disjunction` component. The transformation adds the `Disjunct` and `Disjunction` components within the objects representing the complementarity conditions. It then recasts the modified complementarity components into simple `Block` components. This localizes all changes to the model to the individual complementarity components. Subsequent transformation of the disjunctive expressions to algebraic constraints can be effected through either Big-M (`gdp.bigm`) or Convex Hull (`gdp.chull`) transformations.

13.3.4 AMPL Solver Interface

Solvers like PATH [14] have been tailored to work with the AMPL Solver Library (ASL). AMPL uses `nl` files to communicate with solvers, which read `nl` files with the ASL. Pyomo can also create `nl` files, and the `mpec.nl` transformation processes `Complementarity` components into a canonical form suitable for this format [16].

13.4 Solver Interfaces and Meta-Solvers

Pyomo supports interfaces to third-party solvers as well as meta-solvers that apply transformations and third-party solvers, perhaps in an iterative manner. The pyomo.mpec package includes an interface to the PATH solver, as well as several meta-solvers. These are described in this section, and examples are provided employing the pyomo command-line interface.

13.4.1 Nonlinear Reformulations

The mpec.simple_nonlinear transformation provides a generic way for transforming an MPEC into a nonlinear program. When the MPEC only has continuous decision variables, the resulting model can be optimized by a wide range of solvers.

For example, the pyomo command-line interface allows the user to specify a nonlinear solver and a model transformation applied to a model:

```
pyomo solve --solver=ipopt \
            --transform=mpec.simple_nonlinear ex1a.py
```

This example illustrates the use of the ipopt interior-point solver to solve a problem generated with the mpec.simple_nonlinear transformation. When a transformation is used directly like this, the results returned to the user include decision variables for the transformed model. Pyomo does not have general capabilities for mapping a solution back into the space from the original model. In this example, the results object includes values for the x variables as well as the variables v introduced when applying the transformation to the standard form as shown previously.

Pyomo includes a meta-solver, mpec_nlp that applies the nonlinear transformation, performs optimization, and then returns results for the original decision variables. For example, mpec_nlp executes the same logic as the previous pyomo example:

```
pyomo solve --solver=mpec_nlp ex1a.py
```

Additionally, this meta-solver can also manipulate the ε values in the model, starting with larger values and iteratively tightening them to generate a more accurate model.

```
pyomo solve --solver=mpec_nlp \
            --solver-options="epsilon_initial=0.1 \
                epsilon_final=1e-7" \
            ex1a.py
```

This approach may be useful when using a nonlinear solver that has difficulty optimizing with equality constraints.

13.4.2 Disjunctive Reformulations

The `mpec.simple_disjunction` transformation provides a generic way for transforming an MPEC into a disjunctive program. The `mpec_minlp` solver applies this transformation to create a nonlinear disjunctive program, and then further reformulates the disjunctive model using a "Big-M" transformation provided by the `pyomo.gdp` package. The resulting transformation is similar to the reformulation of bilevel models described by Fortuny-Amat and McCarl [20]. If the original model was nonlinear, then the resulting model is a mixed-integer nonlinear program (MINLP). Pyomo includes interfaces to solvers that use the AMPL Solver Library (ASL), so `mpec_minlp` can optimize nonlinear MPECs with a solver like Couenne [10].

If the original model was a linear MPEC, then the resulting model is a mixed-integer linear program able to be globally optimized (e.g., see Hu et al. [33], Júdice [36]). For example, the `pyomo` command can be used to execute the `mpec_minlp` solver using a specified MIP solver:

```
pyomo solve --solver=mpec_minlp \
            --solver-options="solver=glpk" ralph1.py
```

Note that Pyomo includes interfaces to a variety of MIP solvers, including CPLEX, Gurobi, CBC, and GLPK.

13.4.3 PATH and the ASL Solver Interface

Pyomo's solver interface for the AMPL Solver Library (ASL) applies the `mpec.nl` transformation, writes an AMPL `.nl` file, executes an ASL solver, and then loads the solution into the original model. Pyomo provides a custom interface to the PATH solver [14], which simply allows the solver to be specified as `path` while the solver executable is named `pathamp`.

The `pyomo` command can execute the PATH solver by simply specifying the `path` solver name. For example, consider the `munson1` problem from MCPLIB:

```
# munson1.py
import pyomo.environ as pyo
from pyomo.mpec import Complementarity, complements


model = pyo.ConcreteModel()

model.x1 = pyo.Var()
model.x2 = pyo.Var()
model.x3 = pyo.Var()

model.f1 = Complementarity(expr=complements(
                model.x1 >= 0,
                model.x1 + 2*model.x2 + 3*model.x3 >= 1))
```

```
model.f2 = Complementarity(expr=complements(
                model.x2 >= 0,
                model.x2 - model.x3 >= -1))

model.f3 = Complementarity(expr=complements(
                model.x3 >= 0,
                model.x1 + model.x2 >= -1))
```

This problem can be solved with the following command:

```
pyomo solve --solver=path munson1.py
```

13.5 Discussion

Pyomo supports the ability to model complementarity conditions in a manner that is similar to other AMLs. For example, the `pyomo-model-libraries` repository [52] includes Pyomo formulations for many of the MacMPEC [41] and MC-PLIB [12] models, which were originally formulated in GAMS and AMPL. However, Pyomo does not currently support related modeling capabilities for equilibrium models, variational inequalities and embedded models, which are supported by the GAMS extended mathematical programming framework [18].

Appendix A
A Brief Python Tutorial

Abstract This chapter provides a short tutorial of the Python programming language. This chapter briefly covers basic concepts of Python, including variables, expressions, control flow, functions, and classes. The goal is to provide a reference for the Python constructs used in the book. A full introduction to Python is provided by resources such as those listed at the end of the chapter.

A.1 Overview

Python is a powerful programming language that is easy to learn. Python is an interpreted language, so developing and testing Python software does not require the compilation and linking required by traditional software languages like FORTRAN and C. Furthermore, Python includes a command-line interpreter that can be used interactively. This allows the user to work directly with Python data structures, which is invaluable for learning about data structure capabilities and for diagnosing software failures.

Python has an elegant syntax enabling programs to be written in a compact, readable style. Programs written in Python are typically much shorter than equivalent software developed with languages like C, C++, or Java because:

- Python supports many high-level data types that simplify complex operations.
- Python uses indentation to group statements, which enforces a clean coding style.
- Python uses dynamically typed data, so variable and argument declarations are not necessary.

Python is a highly structured programming language providing support for large software applications. Consequently, Python is a much more powerful language than many scripting tools (e.g., shell languages and Windows batch files). Python also includes modern programming language features like object-oriented programming,

M. L. Bynum et al., *Pyomo — Optimization Modeling in Python*, Springer Optimization and Its Applications 67, https://doi.org/10.1007/978-3-030-68928-5_14

as well as a rich set of built-in standard libraries avaibable for use to quickly build sophisticated software applications.

The goal in this Appendix is to provide a reference for Python constructs used in the rest of the book. A full introduction to Python is provided by resources such as those listed at the end of the chapter.

A.2 Installing and Running Python

Python codes are executed using an interpreter. When this interpreter starts, a command prompt is printed and the interpreter waits for the user to enter Python commands. For example, a standard way to get started with Python is to execute the interpreter from a shell environment and then print "Hello World":

```
% python
Python 3.7.4 (default, Aug 13 2020, 20:35:49)
[GCC 7.3.0] :: Anaconda, Inc. on linux
Type "help", "copyright", "credits" or "license" for more \
    information.
>>> print ("Hello World")
Hello World
>>>
```

On Windows the `python` command can be launched from the DOS shell (or other shells), and on *nix (which includes Macs) the `python` command can be launched from a bash or csh shell (or terminal). The Python interactive shell is similar to these shell environments; when a user enters a valid Python statement, it is immediately evaluated and its corresponding output is immediately printed.

The interactive shell is useful for interrogating the state of complex data types. In most cases, this will involve single-line statements, like the `print` function shown in the previous example. Multi-line statements can also be entered into the interactive shell. Python uses the "..." prompt to indicate that a continuation line is needed to define a valid multi-line statement. For example, a conditional statement requires a block of statements defined on continuation lines:

```
>>> x = True
>>> if x:
...     print("x is True")
... else:
...     print("x is False")
...
x is True
```

NOTE: Proper indentation is required for multi-line statements executed in the interactive shell.

> **NOTE:** True is a predefined Python literal so x = True assigns this value to x in the same way the predefined literal 6 would be assigned by x = 6.

The Python interpreter can also be used to execute Python statements in a file, which allows the automated execution of complex Python programs. Python source files are text files, and the convention is to name source files with the `.py` suffix. For example, consider the `example.py` file:

```
# This is a comment line, which is ignored by Python

print("Hello World")
```

The code can be executed in several ways. Perhaps the most common is to execute the Python interpreter within a shell environment:

```
% python example.py
Hello World
%
```

On Windows, Python programs can be executed by double clicking on a `.py` file; this launches a console window in which the Python interpreter is executed. The console window terminates immediately after the interpreter executes, but this example can be easily adapted to wait for user input before terminating:

```
# A modified example.py program
print("Hello World")

import sys
sys.stdin.readline()
```

A.3 Python Line Format

Python does not make use of begin-end symbols for blocks of code. Instead, a colon is used to indicate the end of a statement defining the start of a block and then indentation is used to demarcate the block. For example, consider a file containing the following Python commands:

```
# This comment is the first line of LineExample.py
# all characters on a line after the #-character are
# ignored by Python

print("Hello World, how many people do you have?")

population = 400

if population > 300:
    print("Wow!")
    print("That's a lot of people")
else:
```

```
print("That's fewer than I suspected...")
```

When passed to Python, this program will cause some text to be output.

Because indentation has meaning, Python requires consistency. The following program will generate an error message because the indentation within the if-block is inconsistent:

```
# This comment is the first line of BadIndent.py,
# which will cause Python to give an error message
# concerning indentation.

print("Hello World, what's happening?")

ans = "A lot"

if ans == "A lot":
    print("Very interesting")
        print("But not without risks.")
else:
    print("Take it easy!")
```

Generally, each line of a Python script or program contains a single statement. Long statements with long variable names can result in very long lines in a Python script. Although this is syntactically correct, it is sometimes convenient to split a statement across two or more lines. The backslash (\) tells Python that text that is logically part of the current line will be continued on the next line. In a few situations, a backslash is not needed for line continuation. For Pyomo users, the most important case where the backslash is not needed is in the argument list of a function. Arguments to a function are separated by commas, and after a comma the arguments can be continued on the next line without using a backslash.

Conversely, it is sometimes possible to combine multiple Python statements on one line. However, we recommend against it as a matter of style and to enhance maintainability of code.

A.4 Variables and Data Types

Python variables do not need to be explicitly declared. A statement assigning a value to an undefined symbol implicitly declares the variable. Additionally, a variable's type is determined by the data it contains. The statement

```
population=200
```

creates a variable called `population`, and it has the integer type because `200` is an integer. Python is case sensitive, so the statement

```
Population = "More than yesterday."
```

creates a variable that is not the same as `population`. The assignment

```
population = Population
```

would cause the variable population to have the same value as Population and therefore the same type.

A.5 Data Structures

This section summarizes Python data structures helpful in scripting Pyomo applications. Many Python and Pyomo data structures can be accessed by indexing their elements. Pyomo typically starts indices and ranges with a one, while Python is zero based.

A.5.1 Strings

String literals can be enclosed in either single or double quotes, which enables the other to be easily included in a string. Python supports a wide range of string functions and operations. For example, the addition operator (+) concatenates strings. To cast another type to string, use the str function. The Python line:

```
NameAge = "SPAM was introduced in " + str(1937)
```

assigns a string to the Python variable called NameAge.

A.5.2 Lists

Python lists correspond roughly to arrays in many other programming languages. Lists can be accessed element by element, as an entire list, or as a partial list. The slicing character is a colon (:) and negative indices indicate indexing from the end. The following Python session illustrates these operations:

```
>>> a = [3.14, 2.72, 100, 1234]
>>> a
[3.14, 2.72, 100, 1234]
>>> a[0]
3.14
>>> a[-2]
100
>>> a[1:-1]
[2.72, 100]
>>> a[:2] + ['bacon', 2*2]
[3.14, 2.72, 'bacon', 4]
```

The addition operator concatenates lists, and multiplication by an integer replicates lists.

> **NOTE:** In Python, lists can have mixed types such as the mixture of floats and integers just given.

There are many list functions. Perhaps the most common is the `append` function, which adds elements to the end of a list:

```
>>> a = []
>>> a.append(16)
>>> a.append(22.4)
>>> a
[16, 22.4]
```

A.5.3 Tuples

Tuples are similar to lists, but are intended to describe multi-dimensional objects. For example, it would be reasonable to have a list of tuples. Tuples differ from lists in that they use parentheses rather than square brackets for initialization. Additionally, the members of a list can be changed by assignment while tuples cannot be changed (i.e., tuples are immutable while lists are mutable). Although parentheses are used for initialization, square brackets are used to reference individual elements in a tuple (which is the same as for lists; this allows members of a list of tuples to be accessed with code that looks like access to an array).

Suppose we have a tuple intended to represent the location of a point in a three-dimensional space. The origin would be given by the tuple $(0,0,0)$. Consider the following Python session:

```
>>> orig = (0,0,0)
>>> pt = (-1.1, 9, 6)
>>> pt[1]
9
>>> pt = orig
>>> pt
(0, 0, 0)
```

For example, the statement

```
pt[1] = 4
```

would generate an error because tuple elements cannot be overwritten. Of course the entire tuple can be overwritten, since the assignment only impacts the variable containing the tuple.

A.5.4 Sets

Python sets are extremely similar to Pyomo Set components. Python sets cannot have duplicate members and are unordered. They are declared using the `set` function, which takes a list (perhaps an empty list) as an argument. Once a set has been created, it has member functions for operations such as `add` (one new member), `update` (with multiple new members), and `discard` (existing members). The following Python session illustrates the functionality of `set` objects:

```
>>> A = set([1, 3])
>>> B = set([2, 4, 6])
>>> A.add(7)
>>> C = A | B
>>> print(C)
set([1, 2, 3, 4, 6, 7])
```

NOTE: Lowercase `set` refers to the built-in *Python* object. Uppercase `Set` refers to the *Pyomo* component.

A.5.5 Dictionaries

Python dictionaries are somewhat similar to lists; however, they are unordered and they can be indexed by any immutable type (e.g., strings, numbers, tuples composed of strings and/or numbers, and more complex objects). The indices are called keys, and within any particular dictionary the keys must be unique. Dictionaries are created using curly brackets, and they can be initialized with a list of key-value pairs separated by commas. Dictionary members can be added by assignment of a value to the dictionary key. The values in the dictionary can be any object (even other dictionaries), but we will restrict our attention to simpler dictionaries. Here is an example:

```
>>> D = {'Bob':'123-1134',}
>>> D['Alice'] = '331-9987'
>>> print(D)
{'Bob': '123-1134', 'Alice': '331-9987'}
>>> print(D.keys())
['Bob', 'Alice']
>>> print(D['Bob'])
123-1134
```

A.6 Conditionals

Python supports conditional code execution using structures like:

```
if CONDITIONAL1:
    statements
elif CONDITIONAL2:
    statements
else:
    statements
```

The `elif` and `else` statements are optional and any number of `elif` statements can be used. Each conditional code block can contain an arbitrary number of statements. The conditionals can be replaced by a logical expression, a call to a boolean function, or a boolean variable (and it could even be called `CONDITIONAL1`). The boolean literals `True` and `False` are sometimes used in these expressions. The following program illustrates some of these ideas:

```
x = 6
y = False

if x == 5:
    print("x happens to be 5")
    print("for what that is worth")
elif y:
    print("x is not 5, but at least y is True")
else:
    print("This program cannot tell us much.")
```

A.7 Iterations and Looping

As is typical for modern programming languages, Python offers `for` and `while` looping as modified by `continue` and `break` statements. When an `else` statement is given for a `for` or `while` loop, the code block controlled by the `else` statement is executed when the loop terminates. The `continue` statement causes the current block of code to terminate and transfers control to the loop statement. The `break` command causes an exit from the entire looping construct.

The following example illustrates these constructs:

```
D = {'Mary':231}
D['Bob'] = 123
D['Alice'] = 331
D['Ted'] = 987

for i in sorted(D):
    if i == 'Alice':
        continue
    if i == 'John':
        print("Loop ends. Cleese alert!")
```

```
      break;
   print(i+" "+str(D[i]))
else:
   print("Cleese is not in the list.")
```

In this example, the for-loop iterates over all keys in the dictionary. The `in` keyword is particularly useful in Python to facilitate looping over iterable types such as lists and dictionaries. Note that the order of keys is arbitrary; the `sorted()` function can be used to sort them.

This program will print the list of keys and dictionary entries, except for the key "Alice," and then it prints "Cleese is not in the list." If the name "John" was one of the keys, the loop would terminate whenever it was encountered and in that case, the `else` clause would skipped because `break` causes control to exit the entire looping structure, including its `else`.

A.8 Generators and List Comprehensions

Generators and list comprehensions are closely related. List comprehensions are commonly used in Pyomo models because they create a list "on-the-fly" using a concise syntax. Generators allow iteration over a list without creating it.

Before discussing list comprehensions and generators, it is helpful to review the Python `range` function. It takes up to three integer arguments: `start`, `beyond`, and `step`. The `range` function returns a list beginning with `start`, adds `step` to it for each element, and stops without creating `beyond`. The default value for `start` is zero and the default value for `step` is one. If only one argument is given, it is `beyond`. If two arguments are given, then they are `start` and `beyond`.

A list comprehension is an expression within square brackets specifying the creation of a list. The following Python session illustrates the use of a list comprehension generating the squares of the first five natural numbers:

```
>>> a = [i*i for i in range(1,6)]
>>> a
[1, 4, 9, 16, 25]
```

Generators are used in iteration expressions in a fashion similar to list comprehensions, but they do not actually create a list. In some situations, the memory and time savings resulting from using a generator versus a list comprehension can be important.

A.9 Functions

Python functions can take objects as arguments and return objects. Because Python offers built-in types like tuples, lists, and dictionaries, it is easy for a function to return multiple values in an orderly way. Writers of a function can provide default values for unspecified arguments, so it is common to have Python functions that can be called with a variable number of arguments. In Python, a function is also an object; consequently, functions can be passed as arguments to other functions.

Function arguments are passed by reference, but many types in Python are immutable so it can be a little confusing for new programmers to determine which types of arguments can be changed by a function. It is somewhat uncommon for Python developers to write functions to make changes to the values of any of their arguments. However, if a function is a member of a class, it is very common for the function to change data within the object calling it.

User-defined functions are declared with a `def` statement. The `return` statement causes the function to end and the specified values to be returned. There is no requirement that a function return anything; the end of the function's indent block can also signal the end of a function. Some of these concepts are illustrated by the following example:

```python
def Apply(f, a):
    r = []
    for i in range(len(a)):
        r.append(f(a[i]))
    return r

def SqifOdd(x):
    # if x is odd, 2*int(x/2) is not x
    # due to integer divide of x/2
    if 2*int(x/2) == x:
        return x
    else:
        return x*x

ShortList = range(4)
B = Apply(SqifOdd, ShortList)
print(B)
```

This program prints [0, 1, 2, 9]. The `Apply` function assumes it has been passed a function and a list; it builds up a new list by applying the function to the list and then returns the new list. The `SqifOdd` function returns its argument (x) unless `2*int(x/2)` is not x. If x is an odd integer, then `int(x/2)` will truncate `x/2` so two times the result will not be equal to x.

A somewhat advanced programming topic is the writing and use of function wrappers. There are multiple ways to write and use wrappers in Python, but we will now briefly introduce *decorators* because they are sometimes used in Pyomo models and scripts. Although the definition of a decorator can be complicated, the use of one is simple: an at-sign followed by the name of the decorator is placed on the line above the declaration of the function to be decorated.

Next is an example of the definition and use of a silly decorator to change 'c' to 'b' in the return values of a function.

```python
# An example of a silly decorator to change 'c' to 'b'
# in the return value of a function.
```

```
def ctob_decorate(func):
  def func_wrapper(*args, **kwargs):
     retval = func(*args, **kwargs).replace('c','b')
     return retval.replace('C','B')
  return func_wrapper

@ctob_decorate
def Last_Words():
  return "Flying Circus"

print(Last_Words()) # prints: Flying Birbus
```

In the definition of the decorator, whose name is `ctob_decorate`, the function wrapper, whose name is `func_wrapper` uses a fairly standard Python mechanism for allowing arbitrary arguments. The function passed in to the formal argument called `func` is assumed by the wrapper to return a string (this is not checked by the wrapper). Once defined, the wrapper can then be used to decorate any number of functions. In this example, the function `Last_Words` is decorated, which has the effect of modifying its return value.

A.10 Objects and Classes

Classes define objects. Put another way: objects instantiate classes. Objects can have members that are data or functions. In this context, functions are often called methods. As an aside, we note that in Python both data and functions are technically objects, so it would be correct to simply say objects can have member objects.

User-defined classes are declared using the `class` command and everything in the indent block of a class command is part of the class definition. An overly simple example of a class is a storage container printing its value:

```
class IntLocker:
  sint = None
  def __init__(self, i):
     self.set_value(i)
  def set_value(self, i):
     if type(i) is not int:
        print("Error: %d is not integer." % i)
     else:
        self.sint = i
  def pprint(self):
     print("The Int Locker has "+str(self.sint))

a = IntLocker(3)
a.pprint() # prints: The Int Locker has 3
a.set_value(5)
a.pprint() # prints: The Int Locker has 5
```

The class `IntLocker` has a member data element called `sint` and two member functions. When a member function is called, Python automatically supplies the object as the first argument. Thus, it makes sense to list the first argument of a member function as `self`, because this is the way a class can refer to itself. The `__init__` method is a special member function automatically called when an object is created; this function is not required.

New attributes can easily be attached to a Python object. For example, for an object named `a` one can attach an attribute called `name` with the value "spam" using:

```
a.name = "spam"
```

It is also possible to query objects to see what attributes they already have.

A.11 Assignment, `copy` and `deepcopy`

A.11.1 References

An assignment statement in Python associates a reference with the variable name given on the left-hand side of the equals sign. If the value on the right-hand side is a literal, then Python creates that thing in memory and assigns a reference to it. For example, consider the following Python session:

```
>>> x = [1,2,3]
>>> y = x
>>> x[0] = 3
>>> x.append(6)
>>> print(y)
```

will result in the output

```
[3,2,3,6]
```

But a subtle point is `y` references the same thing `x` references, not `x` itself. So to continue the example:

```
>>> x = [1,2,3]
>>> y = x
>>> x[0] = 3
>>> x = ''Norwegian Blue''
>>> print(x, y)
```

will result in the output

```
Norwegian Blue [3,2,3]
```

A few types in Python are *immutable*, which means their value in memory cannot be changed; among them are int, float, decimal, bool, string, and tuple. Apart from tuple, this is not surprising. Consider the following:

```
>>> x = (1,2,3)
>>> y = x
>>> x[0] = 3
```

This Python session will result in an error because tuples (unlike lists) cannot be changed once they are created.

A.11.2 Copying

Sometimes assignment of a reference is not what is wanted. For these situations, the Python `copy` module allows for transfer of data from one variable to another. It supports *shallow* copies via the `copy` method and *deep* copies via the `deepcopy` method. The difference is only apparent for compound structures (e.g. a dictionary containing lists). The `deepcopy` method will attempt to make a new copy of everything, while `copy` will only make a new copy of the top level and will try to create references for everything else. For example,

```
>>> import copy
>>> x = [1,2,3]
>>> y = copy.deepcopy(x)
>>> x[0] = 3
>>> x.append(6)
>>> print(x,y)
```

will result in the output

```
[3,2,3,6]  [1,2,3]
```

In this particular example `copy.copy` and `copy.deepcopy` would have the same behavior.

A.12 Modules

A *module* is a file containing Python statements. For example, any file containing a Python "program" defining classes or functions is a module. Definitions from one module can be made available in another module (or program file) via the `import` command, which can specify which names to import or specify the import of all names by using an asterisk.

Python is typically installed with many standard modules present, such as `types`. The command `from types import *` causes the import of all names from the `types` module.

Multiple module files in a directory can be organized into a *package*, and packages can contain modules and sub-packages. Imports from a package can use a statement giving the package name (i.e., directory name) followed by a dot followed by a the module name. For example, the command

```
import pyomo.environ as pyo
```

imports the `environ` module from the `pyomo` package and makes the names in this module available through the local name `pyo`. Analogous to the `__init__` method in a Python class, an `__init__.py` file can be included in a directory and any code therein is executed when that module is imported.

A.13 Python Resources

- *Python Home Page*, `http://www.python.org`.
- *stack overflow*, `https://stackoverflow.com`

Bibliography

[1] AIMMS. Home page. http://www.aimms.com, 2017.

[2] AMPL. Home page. http://www.ampl.com, 2017.

[3] P. Anbalagan and M. Vouk. On reliability analysis of open source software - FEDORA. In *19th International Symposium on Software Reliability Engineering*, 2008.

[4] APLEpy. APLEpy: An open source algebraic programming language extension for Python. http://aplepy.sourceforge.net, 2005.

[5] S. Bailey, D. Ho, D. Hobson, and S. Busenberg. Population dynamics of deer. *Mathematical Modelling*, 6(6):487–497, 1985.

[6] E. Balas. Disjunctive programming and a hierarchy of relaxations for discrete optimization problems. *SIAM J Alg Disc Math*, 6(3):466–486, 1985.

[7] B. Bequette. *Process control: modeling, design, and simulation*. Prentice Hall, 2003.

[8] BSD. Open Source Initiative (OSI) - the BSD license. http://www.opensource.org/licenses/bsd-license.php, 2009.

[9] COIN-OR. Home page. http://www.coin-or.org, 2017.

[10] COUENNE. Home page. http://www.coin-or.org/Couenne, 2017.

[11] CPLEX. http://www.cplex.com, July 2010.

[12] S. P. Dirkse and M. C. Ferris. MCPLIB: A collection of nonlinear mixed-complementarity problems. *Optimization Methods and Software*, 5(4):319–345, 1995.

[13] I. Dunning, J. Huchette, and M. Lubin. Jump: A modeling language for mathematical optimization. *SIAM Review*, 59(2):295–320, 2017. doi: 10.1137/15M1020575.

[14] M. C. Ferris and T. S. Munson. Complementarity problems in GAMS and the path solver. *Journal of Economic Dynamics and Control*, 24(2):165–188, 2000.

[15] M. C. Ferris and J. S. Pang. Engineering and economic applications of complementarity problems. *SIAM Review*, 39(4):669–713, 1997.

© The Author(s), under exclusive license to Springer Nature Switzerland AG 2021
M. L. Bynum et al., *Pyomo — Optimization Modeling in Python*, Springer Optimization and Its Applications 67, https://doi.org/10.1007/978-3-030-68928-5

[16] M. C. Ferris, R. Fourer, and D. M. Gay. Expressing complementarity problems in an algebraic modeling language and communicating them to solvers. *SIAM J. Optimization*, 9(4):991–1009, 1999.

[17] M. C. Ferris, S. P. Dirkse, and A. Meeraus. Mathematical programs with equilibrium constraints: Automatic reformulation and solution via constrained optimization. In T. J. Kehoe, T. N. Srinivasan, and J. Whalley, editors, *Frontiers in Applied General Equilibrium Modeling*, pages 67–93. Cambridge University Press, 2005.

[18] M. C. Ferris, S. P. Dirkse, J.-H. Jagla, and A. Meeraus. An extended mathematical programming framework. *Computers and Chemical Engineering*, 33 (12):19731982, 2009.

[19] FLOPC++. Home page. https://projects.coin-or.org/FlopC++, 2017.

[20] J. Fortuny-Amat and B. McCarl. A representation and economic interpretation of a two-level programming problem. *The Journal of the Operations Research Society*, 32(9):783–792, 1981.

[21] R. Fourer, D. M. Gay, and B. W. Kernighan. *AMPL: A Modeling Language for Mathematical Programming, 2nd Ed.* Brooks/Cole–Thomson Learning, Pacific Grove, CA, 2003.

[22] GAMS. Home page. http://www.gams.com, 2008.

[23] D. Gay. Hooking your solver to ampl. *Numerical Analysis Manuscript*, pages 93–10, 1993.

[24] D. Gay. Writing. nl files, 2005.

[25] GLPK. GLPK: GNU linear programming toolkit. http://www.gnu.org/software/glpk, 2009.

[26] H. J. Greenberg. A bibliography for the development of an intelligent mathematical programming system. *ITORMS*, 1(1), 1996.

[27] J. L. Gross and J. Yellen. *Graph Theory and Its Applications, 2nd Edition*. Chapman & Hall/CRC, 2006.

[28] GUROBI. Gurobi optimization. http://www.gurobi.com, July 2010.

[29] P. T. Harker and J. S. Pang. Finite-dimensional variational inequality and nonlinear complementarity problems: A survey of theory, algorithms and applications. *Mathematical Programming*, 48:161–220, 1990.

[30] W. E. Hart, J.-P. Watson, and D. L. Woodruff. Pyomo: Modeling and solving mathematical programs in Python. *Mathematical Programming Computation*, 3:219–260, 2011.

[31] K. K. Haugen, A. Lkketangen, and D. L. Woodruff. Progressive hedging as a meta-heuristic applied to stochastic lot-sizing. *European Journal of Operational Research*, 132(1):116 – 122, 2001.

[32] A. Holder, editor. *Mathematical Programming Glossary*. INFORMS Computing Society, http://glossary.computing.society.informs.org, 2006–11. Originally authored by Harvey J. Greenberg, 1999-2006.

[33] J. Hu, J. E. Mitchell, J.-S. Pang, K. P. Bennett, and G. Kunapuli. On the global solution of linear programs with linear complementarity constraints. *SIAM J. Optimization*, 19(1):445–471, 2008.

[34] Ipopt. Home page. https://projects.coin-or.org/Ipopt, 2017.

[35] D. Jacobson and M. Lele. A transformation technique for optimal control problems with a state variable inequality constraint. *Automatic Control, IEEE Transactions on*, 14(5):457–464, Oct 1969.

[36] J. J. Júdice. Algorithms for linear programming with linear complementarity constraints. *TOP*, 20(1):4–25, 2011.

[37] J. Kallrath. *Modeling Languages in Mathematical Optimization*. Kluwer Academic Publishers, 2004.

[38] S. Lee and I. E. Grossmann. New algorithms for nonlinear generalized disjunctive programming. *Comp.Chem.Engng*, 24(9-10):2125–2141, 2000.

[39] J. Löfberg. YALMIP: A toolbox for modeling and optimization in MATLAB. In *2004 IEEE Intl Symp on Computer Aided Control Systems Design*, 2004.

[40] Z.-Q. Lou, J.-S. Pang, and D. Ralph. *Mathematical Programming with Equilibrium Constraints*. Cambridge University Press, Cambridge, UK, 1996.

[41] MacMPEC. MacMPEC: AMPL collection of MPECs. https://wiki.mcs.anl.gov/leyffer/index.php/MacMPEC, 2000.

[42] MATLAB. *User's Guide*. The MathWorks, Inc., 1992.

[43] T. S. Munson. *Algorithms and Environments for Complementarity*. PhD thesis, University of Wisconsin, Madison, 2000.

[44] B. Nicholson, J. D. Siirola, J.-P. Watson, V. M. Zavala, and L. T. Biegler. pyomo.dae: A modeling and automatic discretization framework for optimization with differential and algebraic equations. *Mathematical Programming Computation*, 2018.

[45] J. Nocedal and S. Wright. Numerical optimization, series in operations research and financial engineering, 2006.

[46] OpenOpt. Home page. https://pypi.python.org/pypi/openopt, 2017.

[47] OptimJ. Wikipedia page. https://en.wikipedia.org/wiki/OptimJ, 2017.

[48] J. Outrata, M. Kocvara, and J. Zowe. *Nonsmooth Approach to Optimization Problems with Equilibrium Constraints*. Kluwer Academic Publishers, Dordrecht, 1998.

[49] PuLP. A Python linear programming modeler. https://pythonhosted.org/PuLP/, 2017.

[50] PyGlpk. PyGlpk: A Python module which encapsulates GLPK. http://www.tfinley.net/software/pyglpk, 2011.

[51] Pyipopt. Home page. https://github.com/xuy/pyipopt, 2017.

[52] pyomo-model-libraries. Models and examples for Pyomo. https://github.com/Pyomo/pyomo-model-libraries, 2015.

[53] Pyomo Software. Github site. https://github.com/Pyomo, 2017.

[54] R. Raman and I. E. Grossmann. Modelling and computational techniques for logic based integer programming. *Comp.Chem.Engng*, 18(7):563–578, 1994.

[55] N. Sawaya and I. E. Grossmann. Computational implementation of non-linear convex hull reformulation. *Comp.Chem.Engng*, 31(7):856–866, 2007.

[56] H. Schichl. Models and the history of modeling. In J. Kallrath, editor, *Modeling Languages in Mathematical Optimization*, Dordrecht, Netherlands, 2004. Kluwer Academic Publishers.

[57] TOMLAB. TOMLAB optimization environment. `http://www.tomopt.com/tomlab`, 2008.

[58] H. P. Williams. *Model Building in Mathematical Programming*. John Wiley & Sons, Ltd., fifth edition, 2013.

[59] Y. Zhou and J. Davis. Open source software reliability model: An empirical approach. *ACM SIGSOFT Software Engineering Notes*, 30:1–6, 2005.

Index

Printed in the United States
by Baker & Taylor Publisher Services